Atom

Piers Bizony

Icon Books

Published in the UK in 2007 by
Icon Books Ltd, The Old Dairy,
Brook Road, Thriplow,
Cambridge SG8 7RG
email: info@iconbooks.co.uk
www.iconbooks.co.uk

Sold in the UK, Europe, South Africa and Asia
by Faber & Faber Ltd, 3 Queen Square,
London WC1N 3AU
or their agents

Distributed in the UK, Europe, South Africa and Asia
by TBS Ltd, TBS Distribution Centre, Colchester Road
Frating Green, Colchester CO7 7DW

This edition published in Australia in 2007
by Allen & Unwin Pty Ltd,
PO Box 8500, 83 Alexander Street,
Crows Nest, NSW 2065

Distributed in Canada by
Penguin Books Canada,
90 Eglinton Avenue East, Suite 700,
Toronto, Ontario M4P 2YE

ISBN-10: 1-84046-800-9
ISBN-13: 978-1840468-00-7

Typesetting by Hands Fotoset

Contents

List of illustrations

Figures in the text

Pictures in the plate section

Foreword

Now that we have comfortably settled in and acclimatised ourselves to the new millennium, it is tempting to look back at the science of the last century with haughty arrogance at the naivety and innocence of the early pioneers of the atom. Surely we are dealing with loftier, more exciting and more profound issues in science today: from genetic engineering to the challenge of tackling climate change; from the internet to nanotechnology; from artificial intelligence to dark matter, string theory and parallel universes? Sexy science is always that which pushes at the frontiers of our understanding – the questions we have yet to answer and the phenomena we have yet to explain. Why would we wish to dwell on unfashionable stories relegated to the ignominy of a thousand textbooks when we have a brave new world to wonder at and make sense of? What is there left unsaid about those brave pioneers of what is still nostalgically referred to as 'modern physics', beyond the respectful acknowledgements of centenaries of their births and deaths?

I am not a historian of science, nor am I immune to the lure of current ideas and unresolved puzzles in fundamental physics. After all, I spend an unhealthy fraction of my waking hours pondering such things. That is my job. But I put it to you that it is hard to imagine an age of more exciting revelations and revolutions in humanity's quest to understand the ultimate building blocks of our universe than the one that faced the scientists of the first half of the 20th century. Never before or since has mankind successfully unlocked so many of nature's inner secrets. The story of how a band of dedicated men and women slowly but surely unravelled the inner workings of the atom is not only a great drama, but a lesson in scientific collaboration, dedication and resourcefulness that is worth telling to each new generation.

In little over half a century, our understanding of the world developed from debate over whether atoms even existed at all to the realisation that they are indeed the building blocks of the universe, and on to probing their inner workings and structure. It is quite remarkable that we can today describe the rich variety of everything we see around us – from tables and chairs to the ground we stand on and the air we breathe, to our own bodies and the minds we use to think about such things – in terms of just three elementary particles much tinier than we can ever hope to see directly: the electron and two types of quarks; just three basic building blocks that make up everything there is.

It is interesting to note that such a reductionist approach, in which everything in the universe, however complex, can be ultimately described in terms of its constituent parts, its basic building blocks, is today acknowledged by many scientists to be inadequate. A nice example of this is the most basic property of water: its 'wetness'. This is something that emerges only when we are dealing with trillions of water molecules combined together. We would never expect to predict water's wetness by studying a single water molecule and probing its atomic or subatomic structure. Clearly our knowledge that all matter is composed of atoms is not enough to explain the rich variety of phenomena we see in the world around us. Despite this, mankind's discovery of atoms ranks as being *the* most important in the history of science (followed not so far behind, I would argue, by Darwinian natural selection). The story of the atom is a fascinating one, and I don't mean just to nerdy historians of science. Many of the personalities involved in this epic quest are household names. Who hasn't heard of Curie and Einstein? Others are well known to scientists but may be less so to the general public. Of course, you may well know of Rutherford, Bohr, Schrödinger, Heisenberg and Feynman. But the story of Wolfgang Pauli, the man who laid the foundations of modern chemistry, or that of Fred Hoyle and George Gamow who between them explained how all the atoms were created in the first place, is not so widely known and deserves to be told.

Today, it is hard to imagine just what little knowledge Marie and Pierre Curie started with when they embarked on their quest to explain radioactivity, or what a great leap of intuition Rutherford needed to make to figure out what the inside of an atom looks like. What is even more remarkable is that in cracking the code of the atom, physicists needed to invent a new way of thinking, a mathematical construct far stranger than anything they could have dreamt up – ideas so powerful that they form to this day the most important scientific theory ever produced: quantum mechanics. In their quest for simplicity, these geniuses found that Nature was more remarkable than they could ever have imagined. Old notions of a logical mechanistic universe that could be explained by a few simple rules were overthrown so completely that one prominent physicist, questioning the correctness of a new theory presented to him, wondered not whether the ideas it was based on were crazy, but rather whether they were crazy enough!

Being able to describe the nature of the atom was far from the whole story, of course. Its energy was harnessed and its origins were explained. For instance, there is no idea more romantic or powerful than that we are all made of stardust. Every atom on earth, including our own bodies, was forged in the crucibles of space, being either formed at the birth of the universe, just after the Big Bang, or cooked inside a star billions of years ago.

What must it have been like to be part of that great

adventure, designing simple but ingenious experiments on lab benches across Europe, building fantastic machines in the Cavendish in Cambridge that would smash open the atomic nucleus, or filling blackboards in Copenhagen with symbols and algebra, inventing new mathematics as they went along? This book tells that story.

Jim Al-Khalili
January 2007

Jim Al-Khalili is head of the theoretical nuclear physics group in the Department of Physics at the University of Surrey. He has written several popular science books, including *Quantum: A Guide for the Perplexed*, *Nucleus*, and *Black Holes, Wormholes and Time Machines*. He is a regular guest on 'In Our Time' on Radio 4, and is the presenter of the BBC documentary series *Atom* which ties in with this book.

Acknowledgements

The author would like to express his gratitude to Paul Sen and his colleagues, whose fine BBC series *Atom* has been the inspiration for this book. The series presenter, Dr Jim Al-Khalili, was tirelessly patient answering my questions and sympathetically correcting my scientific and historical mistakes. Any errors that remain in this text are my fault alone. Thanks also to Icon Books for making this project possible. They are a very pleasant team to work with. Likewise, I thank Peter Tallack at Conville and Walsh, who first introduced me to Icon. Above all, I would like to thank my family, Fiona, Alma and Oskar, for giving me such a happy atmosphere in which to work. From now on, I promise not to talk about atoms at meal-times.

About the author

Piers Bizony is a science journalist and space historian who writes for magazines such as *Focus* and *Wired* as well as the *Independent*. His last book was *The Man Who Ran the Moon* (Icon, 2006).

Preface

All our cities have been reduced to dust. All our books have been vaporised in a global firestorm, and every last byte of computerised knowledge has been wiped out by the electromagnetic disturbance of some terrible disaster. Of the six billion humans that existed a few days ago, only a few thousands are still alive. Is this the absolute end of civilisation, or can some of the great intellectual achievements from the last two thousand years be rescued from the dust and ashes?

Suppose that the vanished civilisations of Earth had seen this day coming, and had created dozens of blast-proof tablets of toughened steel in the hope that at least one of them might be discovered by survivors of a future calamity. And suppose that these tablets had space for only one quick sentence of information, perhaps repeated in several different languages, as if on a futuristic Rosetta Stone. What might that sentence be? According to Richard Feynman, one of the 20th century's greatest physicists, humanity could recover pretty much everything it needed to know about the material world from the simple statement:

All things are made of atoms, little particles that move around in perpetual motion, attracting each other when they are a little distance apart, but repelling upon being squeezed into one another.

Our quest for the atom has been one of the greatest scientific adventures in history. Not least of the problems is that an atom is a mind-numbingly small entity. No technical comparison can really help us grasp how small an atom is. It's about a tenth of a millionth of a millimetre across. Can you picture that in your mind's eye? Analogies drawn from our everyday experience also leave us reeling. Let's try one. If you dip a cup into the sea and extract a cupful of water, there are as many atoms in that cup as there are cups of water in all the oceans of the world. No? Let's try one more, this time based on an ever-popular yardstick of scientific communication, a strand of human hair. The width of the finest hair is longer than the span of one million carbon atoms stretched out in a row. Still no good? Don't worry. The atom's tiny size is only one of its multitude of ungraspable qualities. Another challenge to our sensory expectations, based as they are on the solidity of tables and chairs, the lumpen density of lead, the satisfying heft of gold coins in our hand, is this. The atom is mainly just empty space. If all the atoms in your body could be squeezed down to get rid of the empty volume inside them, you would weigh the same, but would be as small as a grain of salt. Almost all of you – almost all of everything in the entire cosmos – is *nothing*. The atom is not a hard-edged 'thing'. It's a ghostly shimmer.

Atoms in their turn are made from subatomic components thousands of times smaller again. The instruments designed for atomic science are made out of atoms, so we are essentially trying to measure entities smaller than the smallest tips of the finest instruments that we can ever create. Atomic science is like investigating a slippery coin while wearing a blindfold and a pair of thick gloves, and trying to find out which side of the coin is 'heads' and which 'tails'. Ordinary fumbling is not sufficient. We have to be unusually clever to find the clues: not with, but *despite* the gloves and the blindfold, for we can never take them off. The gloves and the blindfold represent the limits of our animal senses, which allow us only so much access to the world of the ultimately small.

Yet the basic secrets of the atom were explored nearly a hundred years ago, at a time when scientific instruments consisted of little more than glass tubes, hand-made wooden and brass boxes, and simple electrical circuits wired up to batteries. There were no electron microscopes, no positron emission scanners, no supercomputers and plasma screen read-outs. There was just the human mind, the human imagination. Before the atom could be investigated with instruments, it needed to be *imagined*. After all, how could we find any atoms before we knew what we were looking for?

Today the atom is regularly subjected to multi-billion-dollar experiments in super-laboratories employing thousands of people. The tips of our instruments have become so fine that we can move individual atoms from place to place. We can even write words by nudging atoms across super-polished sheets of metal. Yet when we read this writing with our fantastic machines, we still never really see the atoms themselves. What we see are shadowy images of them rendered as computer graphics, which in turn are based on predictions of science that are far from absolutely certain. At heart the atom remains a thing of the mind. We still do not know the reality of it, and this book explains why its mysteries refuse to be resolved.

The atom may not be knowable, but it can be harnessed. With terrifying swiftness we have learned how to liberate its dangerous powers, and now that these inner demons have been released into the world, we cannot expect any time soon to coax them back into their cages. We have also discovered that the world of the unimaginably small is the direct cousin of the unimaginably vast. In a single atom lie all the forces and energies that brought the cosmos into being.

Is this not a fantastic drama? Science is so often seen as coldly objective, and perhaps that's why so many people think of scientists as dull, colourless technicians. But science is a profoundly creative act, and the larger-than-life personalities involved in the quest for the atom were more creative than most. They came up with revolutionary theories that were shaped by their characters as much as by the data from their experiments. Some saw the atom as beyond our reach, an abstraction far removed from the treacherous human realm of tactile experience. Others with a more sensual outlook on life were sure that there must be something real in the atom's shadow. The great figures of atomic discovery in the 20th century could not necessarily tell us what was true about their work. They told us what their favoured *versions* of that truth happened to be. Their joint legacy is still the cause of vigorous argument today, for it calls into question our most cherished concepts of reality.

PART ONE

Energy in Pieces

In the year 1900 a deeply conservative physicist called Max Planck concluded, somewhat reluctantly, that energy is not smooth and continuous. It is divided into discrete amounts, mysterious packets, which he called 'quanta'. It was a discovery that would revolutionise all of science.

The word 'atom' is derived from a Greek idea formulated 2,400 years ago. The philosopher Democritus argued that matter is made from indivisible, imperishable and unchanging particles, which he called *atomos*. It was a brilliant insight, based on logical argument, but Democritus did not choose to test any of his concepts experimentally. Like most of his contemporaries, he thought that logic alone should be able to resolve the mysteries of nature. During the next 22 centuries the atom made almost no impact on the human imagination, until a precocious young teacher at the dawn of the Industrial Revolution discerned the hard-edged practical value of talking about atoms. John Dalton was born in 1766 into a modest Quaker family in Cumberland, and earned his living for most of his life as a teacher, first at his local village school (where he began giving classes at the age of twelve) and then in the factory-dominated city of Manchester. Here he reanimated the atomic theory in a strict mathematical framework rather than just as a vague philosophical idea. He concluded that all atoms of a given element must be identical to each other, and argued that chemical compounds are formed by a combination of two or more different kinds of atoms. Carefully weighing his chemicals before and after they reacted with each other, he worked out the ratios of different elements that went into certain well-known compounds. The atom emerged from his work as a spectacularly reliable chemical counting unit. As a statistical way of looking at gas and steam pressures, the atom was also invaluable. Yet it remained, for now, just that: a workable counting tool with no proper physical explanation behind it.

By the end of the 19th century, a self-confident set of rules had been assembled to describe just about everything that could be looked at, listened to, weighed or measured: the precise movements of the stars and planets across the sky, the temperatures and pressures of gases under given conditions, the rate of transfer of heat from one substance into another, the equations for shaping glass lenses so that they would bring rays of light to a focus, and so on. It was a mechanistic and results-driven way of looking at the world, and it ushered in the age of the electric lightbulb, the radio telegraph, the telephone, the motion picture, the motor car and the aeroplane. Science also seemed capable of unveiling secrets of nature at the profoundest levels. In the 1820s, a number of astronomers insisted that the gravity of an unknown planet must be responsible for observed irregularities in the orbit of an already familiar planet, Uranus. For three decades the pure and logical rule of Newtonian mathematics was their only guide. And then in 1846 they steered their telescopes to the point in the sky where the mathematics said that the planet should be. And there it was. Neptune existed because the classical laws of nature said it had to exist.

Scientists were satisfied of two things. First, almost everything that could be understood *was* understood; and second, the remaining mysteries were the province of religion and metaphysics, not science. Many properties of the commonly available chemical elements – hydrogen, oxygen, carbon, nitrogen, copper, iron, and so on – were predictable. Thanks to Dalton and his successors, chemists knew the precise ratios of elements in thousands of industrially useful compounds. There was an elaborate counting system based on the atom, which was widely held to be the most fundamental unit of all matter. The atom was a useful idea, but it was too small to be seen in any microscope, so its existence could not be verified.

We knew everything. And yet we knew nothing. Scientists were thoroughly accustomed to putting different kinds of knowledge into separate compartments. Botany was popular in the late Victorian age as a respectable hobby for the leisured classes, as was mathematics and the study of optics. The smellier and more hands-on business of chemistry and mechanical engineering were best left to the newly powerful industrialists. Few people would have imagined that all these disciplines might share common ground. No one suspected that the shapes of molecules, the specific arrangements of atoms in chemical compounds, might give strength to a tree, while another arrangement lent flexibility to its rubbery sap, and yet another controlled the shape of its leaves. There were many individual natural philosophers and amateur scientists intrigued by such questions, but the burgeoning numbers of university specialists, military research arsenals and commercial laboratories had no common framework with which to tackle deep, abstract problems in science. There were no grants available to study questions that did not already appear on the list of approved and potentially profitable questions: how could catalytic converters be made more efficient? How could the casings of steam engines be machined in stronger but more lightweight configurations? What mix of explosives could most effectively hurl a 15-pound shell the greatest distance?

The 'what happens when …' questions of science were incredibly well understood by 1900. The 'why' questions were barely addressed. No one had the tools to understand the strange invisible energies emanating from Henri Becquerel's experiments with uranium salts. It was a puzzle, also, that his compounds emitted their energies week after week, month after month, without apparently depleting like any normal energy source. Similarly, and on a grander scale, it was a mystery how the sun could keep shining and not burn itself out. There was no unifying concept that could link all these disparate wonders together and explain them. We didn't even now why red things look red.

A seemingly unremarkable young clerk in the Patent Office at Berne, Switzerland, pondered that the science of his day was like a vast library of unrelated books in myriad subjects. Yet he had faith in 'a definite plan in the arrangement of the books, a mysterious order, which we do not comprehend, but dimly suspect'. The clerk believed that if only we could study how that order worked, instead of concentrating just on the individual books, then the entire library might one day resolve itself into a single, compact and breathtakingly tidy volume. The clerk was not an experimenter and had no laboratory. He drew his conclusions purely from the logic of scientific papers already available to his generation. In 1905, aged just 25, he published three papers which, in principle, should have revolutionised all scientific thinking. A few scientists adopted his ideas enthusiastically, and the clerk quickly gained acceptance in academia, but the world as a whole remained unmoved, and he graduated from youthful revolutionary to middle-aged professor without attracting much attention outside his close-knit scientific coterie. Albert Einstein was 40 years old by the time he even began to become famous.

It often happens in science that an observation in one field of research eventually throws a startling new light on another area. An observation in botany unexpectedly laid the foundations for one of the most important discoveries in all of modern science: the discovery of the atom. And it was Einstein who spotted the clues. His 1905 theory on Special Relativity, for which he is now best known, was only one of several intellectual breakthroughs he published in that extraordinary year. Another paper was all about little grains of pollen.

From botany to the atom

In 1827 the Scottish botanist Robert Brown was examining pollen grains under a simple microscope. He put some grains in a droplet of water, and noticed that they moved about, tracing random zigzag paths across his microscope's field of view. At first, he concluded that the movement of each grain 'arose neither from currents in the fluid, nor from its gradual evaporation, but belonged to the particle itself'. Other botanists enthusiastically decided that Brown had witnessed a fundamental 'life force' animating these tiny pieces of biological matter (typically the pollen grains measured no more than 1/100th of a millimetre across). This was a perfect example of something that tends to happen in science. There can be an unseemly rush to try to confirm theories which have already gained currency among researchers. Observational facts are sometimes interpreted to fit a favourite theory, and this is one of the biggest mistakes that any scientist can make. It's much better to adjust the theory to fit the facts, even – or perhaps especially – when it involves abandoning cherished ideas about how nature works.

Wisely as it turns out, Brown was more cautious. Even as he prepared to publish his results, he revised his text to warn that he had seen a similar motion among pollen grains he had preserved in alcohol many months before, so that surely they must have been lifeless by the time he put them under the microscope. Of course there was a slim chance that pollen was harder to kill than he had assumed, so one more experiment was needed to remove any ambiguity. He ground down some inorganic mineral samples into powders and suspended them in water. Again, he saw random movements through his microscope. If there was some kind of a force at work here, it almost certainly wasn't coming from the grains. And yet they moved. The only logical conclusion was that something in the water was pushing them around.

Throughout the Victorian era, the jiggling of the grains remained an intriguing enigma whose significance was not truly understood. Then, at the dawn of the 20th century, a Swiss-Italian electrical engineer named Michelangelo Besso introduced his very good friend Albert Einstein to what he called 'Brownian motion'. In 1905 Einstein was inspired to write a paper on this theme, in which he described how the motion could be understood as the buffeting of billions of water molecules against the grains. Previous theorists had been confused by the idea that something so infinitesimally small and lightweight as a molecule could push against the comparatively massive grains. Einstein certainly wasn't suggesting that the individual zigs and zags of the grains were caused by each one being hit by individual molecules. They were the cumulative outcomes of many millions of random impacts. He even calculated the probability of the motions in a way that could be tested by subsequent experimenters. Ignoring all the zigs and zags, and focusing only on the straight-line distances covered between the start of a grain's journey and where it ended up after a given amount of time, Einstein accurately predicted how far a grain would travel. In other words he 'smoothed out' the irrelevant details of every last zig and zag, and treated the whole problem statistically.

Yet he drew back from claiming that the apparent action of the molecules on the grains specifically proved the existence of molecules. Instead he suggested that the statistical effects he described would produce the visible large-scale motions observed in Brownian motion *if* it turned out to be the case that molecules existed. As for the molecules themselves, he warned that 'the data available to me are so imprecise that I could not form a judgement on the question.' It was an important and conscientious distinction. For now, these invisible entities remained a useful theoretical model for predicting the thermal and kinetic forces in liquids and gases, and for predicting the outcomes of chemical reactions. But still, no one had yet 'seen' a molecule, let alone an atom.

Smaller than the smallest thing

Strangely enough, someone had by now demonstrated the existence of something even *smaller* than an atom. In 1897 the English physicist Joseph J. Thomson was experimenting with a sealed glass tube from which most of the air had been drawn out. Inside were two metal plates, mounted at opposite ends, and wired up to a battery. The plate attached to the positive terminal was called the anode, and the negative plate was known as a cathode. When the current was switched on, a mysterious glow could be observed at the anode end of the tube. Certain types of glass exhibited a very faint glow unaided, but when a coat of phosphor was applied to the inside of the glass tube at the positive end, the glow was unmistakable. Some kind of invisible beam was crossing the gulf between the cathode and the anode, striking the end walls of the tube and producing the glow.

These 'cathode ray' tubes, tremendously popular among experimenters in the late 1800s, were the ancestors of television. Thomson's special contribution was to prove that electric currents or magnets just outside the vacuum tube could deflect and even steer the rays, altering the positions of the luminous spots on the phosphor screens. A wide variety of different anode and cathode metals produced similar results, so he concluded that the rays were a fundamental constituent of nature, and not just some oddity connected with a particular material. He showed that the rays were narrow beams of negatively charged particles, which he called 'corpuscles'. By measuring the influences of external magnets and electrical currents on these particles, he proved they were 2,000 times less massive than an atom of even the lightest element, hydrogen. We know these particles, today, as electrons.

It was an odd state of affairs, that an entity even smaller than the atom had been identified while the atom itself remained just a theoretical notion. In April 1897 Thomson admitted to a meeting of the Royal Society in London that 'the assumption of a state of matter more finely divided than the atom is a somewhat startling one'. In 1904 he took the bold step of asserting that 'the atom consists of a number of corpuscles [electrons] moving about in a sphere of uniform positive electrification'. This came to be known as the 'plum pudding' model. For now, that's all it was: a mind's eye visualisation, a vague speculation. Yet Thomson's confidence was bolstered by news from Paris, where a brave and romantic couple were at last finding substance in the atom's elusive shadow.

Scientific romance

A hundred years ago, even at the tail end of the Victorian era, a male scientist could get away with having an adulterous affair as long as his work was good enough. A woman had to tread more carefully in her private life, even if her work was in the Nobel Prize-winning class. One who always refused to toe the line was Madame Curie.

Maria Skłodowska was born in 1867 in Warsaw, the fifth and youngest child of Bronisława Boguska, a pianist, singer and teacher, who died of tuberculosis while Maria was still a child, and Władysław Skłodowski, a professor of mathematics and physics. At sixteen Maria won a gold medal for outstanding achievement at her secondary school education. She dreamed now of travelling to Paris and entering the Sorbonne, one of the few major European universities where a young woman might be allowed to study. Unfortunately her father lost all the family savings in a failed investment scheme. Maria was forced to find work as a teacher, and then became a governess, essentially a nanny-cum-tutor role familiar to well-brought-up young women in the Victorian age whose families had fallen on hard times.

Meanwhile, she found time to read Polish books to poor women who would not otherwise have had access to education, except at the hands of Poland's Russian overlords. Maria was a member of a clandestine and politically radical 'flying university' which convened wherever it could, and just as quickly dispersed whenever threatened. The Russian authorities did not approve of Maria's nationalist ideals.

Her next risky adventure was to fall in love with one of the sons of the family she was living with as a governess. The affair was passionate, and the love completely mutual, but the young man's parents refused to let him marry a penniless governess, and Maria was forced to leave the household. Now her only hope was her sister Bronia, who had struck a deal with her some years earlier. Funded by Maria's earnings as a governess, Bronia had made it to Paris, where she was studying for a medical degree. On her return to Poland, the sisters were supposed to swap roles. These two strong-minded women kept to their plan, and in November 1891 Maria at last set out on the thousand-mile train journey to Paris and the Sorbonne. She attended physics and mathematics lectures by day, then at night returned to her very humble student's lodgings in the city's Latin Quarter. She ate little more than bread and butter, and seldom drank anything more costly than tea. Her studies went well, and after three years of dedicated student life, she passed examinations in physics and mathematics with outstanding grades. Marie's goal (she changed her name while in Paris) was to obtain a teacher's diploma and then return to Poland. Her instincts told her that her widowed father must surely expect her to come home and play her part in supporting the family.

And then she chanced to meet her soulmate. 'I was struck by the open expression on his face,' she recalled. 'His simplicity, his smile, at once grave and youthful, inspired confidence.' Thirty-five-year-old Pierre Curie was the head of a laboratory at the School of Industrial Physics and Chemistry. He had already made something of a name for himself by discovering, with help from his brother Jacques, that when an electric current was applied to a quartz crystal, it changed shape by a tiny amount. Conversely, when the crystal was squeezed or pulled, it delivered a jolt of electricity in response. This might sound obscure, yet it was the key to a new range of super-sensitive scientific instruments. Pierre was a skilled designer of measuring equipment, and he was well respected in Paris at that time, although he was not so good at insinuating his way into the professional élite of French science. He despised political games-playing and had little appetite for medals and awards, or any of the other back-slapping perks of his trade.

Marie's good fortune was that this decent and unegotistical man proved very happy to collaborate with her in work as well as in love. She knew she was supposed to return to Poland after completing her studies at the Sorbonne, and she visited her family to give them the news about Pierre, uncertain how they would react. Much to her relief, her father made it quite clear that she should return to Paris straight away and marry him. In July 1895, after an idyllic wedding, one of the most romantic couples in the history of science got down to work, studying the strange invisible rays recently discovered by another French scientist, Henri Becquerel. He had worked with a very unusual element. It was called uranium.

The dollar lode

At the beginning of the 16th century, the rough mountain territory dividing Bohemia from Saxony, the borderland between modern Germany and the Czech Republic, was covered by an impenetrable virgin forest, a refuge for wolves, bears and bandits. The discovery of precious metals triggered the first 'silver rush' in history. The previously insignificant little town of Joachimsthal soon become the largest mining centre in Europe. In just a couple of years an eager influx of chancers swelled the population to 20,000. The silver was minted into a coin called a Joachimsthaler, later known more simply as a thaler. Rather like a certain currency in today's world, the thaler was accepted worldwide. The silver coins are no longer in circulation, but the name has stayed with us, slightly transmuted. It's now pronounced 'dollar'.

After just three decades, Joachimsthal's silver reserves were exhausted. Plagues killed off much of the population, and the Thirty Years War finished the job. Joachimsthal became a ghost town with an unhealthy reputation. Miners had always fallen ill there, even before the plague struck. But just because the silver was gone didn't mean that the mines had nothing left to offer. Along with the silver, the miners had often come across a shiny black mineral, which didn't immediately impress them as being of much use. They called it *Pechblende*, from the German words *Pech*, which means bad luck, and *Blende*, meaning mineral. In 1789 an amateur German chemist, Martin Klaproth, decided to see what it was made of. He found that it contained 'a strange kind of half-metal', which he named in honour of the planet Uranus, at that time believed to be the last planet in the solar system.

During the next century, 'pitchblende' was found in Cornwall, France, Austria and Romania, and by the end of the Victorian era, thousands of scientific papers had been published on geological and mineral occurrences of 'uranium'. The metal, as dense as gold, was apparently the heaviest element on earth. Its principal value appeared to lie in the vivid colours of its oxides and salts, which were used to create glassware with an attractive fluorescent glow, or glazes for ceramics and porcelain in orange, yellow, red, green and black. Some of these techniques had been known since Roman times. No one suspected the invisible dangers lurking inside those decorative flourishes. Uranium's more dramatic potentials emerged in 1896, when Becquerel discovered quite by accident that it gave off invisible rays capable of fogging photographic plates wrapped in light-proof black paper.

Just a few months earlier, the German physicist Wilhelm

Roentgen was experimenting with electron beams in an apparatus similar to J.J. Thomson's cathode ray vacuum tube, except that his electrons streamed from a heated metal cathode. He noticed that a fluorescent phosphor screen at the other end of his workbench began to glow. Yet another mysterious ray had revealed itself, this time by escaping from the tube altogether and flying across the room. When Roentgen placed a thick black card between the tube and the screen, the screen still glowed. Finally, he put his hand in front of the tube, and saw the silhouette of his bones projected onto the screen. He had no way of knowing that the electrons in his vacuum tube, flying off a particularly hot filament, were so energetic that they were generating a different kind of radiation when they knocked into atoms in the glass walls of the tube.

One week later, Roentgen discovered the ray's most beneficial application. He captured an image of his wife's hand on a photographic plate. Her bones (her wedding ring too) showed up as solid shadows. She was horrified by the death-like image, but the medical world quickly embraced this new discovery, which seemed little short of magic. Here was a tool that could look inside a living patient and reveal broken bones, or seek out the exact location of a shard of shrapnel prior to surgery. But what were these rays? An invisible kind of light? Something akin to Thomson's corpuscles? At first no one was sure, and this is why they were called x-rays.

Becquerel's uranium rays seemed to have similar characteristics. Pierre and Marie Curie were fascinated by these exciting emanations, which they described, in French, as 'radio-actif'. The word 'radio', derived from the Latin geometrical term 'radius', did not then have the specific meanings we assign to it today. It simply meant 'related to rays'. The Curies had discovered an invisible 'active ray'. The English derivation of their term is one we all recognise and fear. It is 'radioactive'.

By 1898 they had secured the use of a laboratory space

granted to them by the School of Physics and Chemistry in Paris. Here they began the grimy and labour-intensive business of extracting microscopic quantities of pure uranium from huge, boiling vats of pitchblende solution. In truth, their laboratory was nothing more than a disused dissecting room purloined from a nearby medical school. It had an unreliable glass roof that let the rain in, and it was unbearably hot in summer and inhumanely cold in winter. Visitors were surprised by the roughness of the conditions, and by the noxious smells emanating from the chemical vats and gas burners. It was more like a grim factory than a laboratory. The German chemist Wilhelm Ostwald paid a visit on one of the rare days when the Curies were absent on other business. The lab, he wrote, 'was a cross between a stable and a potato shed, and if I had not seen the worktable and items of chemical apparatus, I would have thought that someone had played a practical joke on me'.

Day after day the routine was the same. Sacks of pitchblende arrived freshly dug from the giant discarded slag heaps that still festered around the abandoned Joachimsthal mines. Marie would clean off the mud and grass and pine needles from the rough lumps, which then had to be ground down into fine powder, then boiled into a liquid which could be sieved and refined yet further. There was an acid bath to dissolve unwanted impurities. Then came an electrolysis process, similar to the method used for silver-plating cutlery. At the end of many months' labour, just a few grammes of purified uranium might be extracted. 'Sometimes I had to spend a whole day mixing a boiling mass with a heavy iron rod nearly as large as myself. I would be broken with fatigue at the day's end', wrote Marie. Yet she had no urge to complain, for Pierre was by her side, and their discoveries were mounting up. 'It was in this miserable old shed that we passed the best and happiest years of our life.'

The odd thing was that the crude pitchblende still seemed more radioactive than the refined uranium. Pierre and Marie realised that there must be other kinds of radioactive element awaiting discovery inside the ore. It took a year for them to extract polonium, which they named in honour 'of the home country of one of us', and radium, which glowed in the dark when suspended as a solution in a jar of water. 'These substances exist in pitchblende only in the form of traces, but they have an enormous radioactivity of an order of magnitude two million times greater than that of uranium', noted Pierre. Marie's brilliant intuition was that radiation did not depend on the type of chemical element under investigation, whether it was radium, polonium, thorium, barium or uranium. The true unit of radioactive emission was the atom.

Where did the internal power supply come from? Pierre speculated that 'radioactive substances borrow from an external radiation the energy which they release. It's not absurd to suppose that space is constantly traversed by very penetrating radiations which certain substances would be capable of capturing in flight.' This was the most comforting and familiar idea. You warm something up, and after a while the warmed object begins to emit heat on its own account. He and Marie knew that this was no ordinary kind of heat energy they were dealing with, even if radium did feel warm to the touch. It was more likely that 'radioactive substances draw from themselves the energy which they release ... The quantity of heat released by radium in several years is enormous if it is compared with the heat released in any chemical reaction with the same weight of matter.'

In 1903 Pierre and Marie shared the Nobel Prize for Physics with Henri Becquerel for their joint discovery of radiation. The prize money certainly helped improve the Curies' straitened financial circumstances. The following year, 45-year-old Pierre was appointed professor at the Sorbonne (a post was created especially for him), and Marie, 37 years old, became his assistant. They were still young enough to hope for more and greater achievements yet to come. And there was more good news. Their healthy six-year-old daughter Irène was joined by a new arrival, Eve.

The worst problems on the horizon seemed to be the endless requests for press interviews, the unwanted fame generated by the Nobel award, and the constant aches, pains and illnesses resulting from the backbreaking work in their lab, not to mention the poor diet forced on them by their meagre wages.

The Curies suspected, but somehow failed to take seriously, that there might be more to their fatigue than this. In his Nobel acceptance speech, Pierre warned that leaving 'a small glass ampoule with several centigrams of a radium salt in one's pocket for a few hours' might produce no ill effects, but several days' exposure would cause 'a redness to appear on the epidermis, and then a sore which will be very difficult to heal. A more prolonged action could lead to paralysis and death. Radium must be transported in a thick box of lead.' He had tested this by strapping a packet of radium to his skin for several hours, experimenting with the possibility that radiations might be used to burn away cancerous growths. Marie liked to keep a little phial of radium by her bedside. It took seven tons of pitchblende to produce a single gramme of the stuff. Little wonder if she enjoyed watching her precious prize glow in the darkness. 'One of our joys was to go into our workroom at night. We then perceived on all sides the feebly luminous silhouettes of the bottles or capsules containing our products. It was a lovely sight and always new to us. The glowing tubes looked like faint fairy lights.'

Suddenly Pierre was killed, and not by radiation, but by the most ordinary kind of accident. On the last day of his life, Thursday, 19 April 1906, he lunched at the Association of Science Professors, then headed off, as scheduled, to go over proofs with his publisher and to visit a nearby library. He and Marie were looking forward to entertaining friends that evening. But when Pierre arrived at his publishers, he found to his annoyance that the doors were closed and no one was in. Hurrying back across the street in pouring rain, he failed to notice a horse-drawn wagon with a load of military stores, weighing some six tons. He was crushed under the wheels of the cart and killed instantly.

Grief-stricken, Marie put all her energy into continuing the work they had begun together, eventually becoming head of a laboratory at the Sorbonne and the first woman lecturer at the university. She comforted herself by writing a journal to Pierre as if he were still in a position to read it. 'I walk as though hypnotised, without a care about anything. I will not kill myself. I don't even have the desire for suicide. But among all those carriages, isn't there one which will make me share the fate of my beloved?' Her students at the Sorbonne found her impressive and even beguiling. One recalled that 'she looked very pale, her face was impassive, her black dress extremely simple. One saw only her luminous large forehead, crowned by abundant ashen hair, which she pulled back tight without succeeding in hiding her beauty.' In 1908 Marie was appointed as a full professor at the Sorbonne, and then received an unprecedented second Nobel Prize in 1911, this time for chemistry, in honour of her discovery of radium. During the First World War, she dedicated herself entirely to the cause of mobile x-ray units, which must have helped save the lives of countless wounded soldiers.

Marie's officially honoured achievements are distinct from her story as a woman. Unlike many respectable widows of her time, she wasn't prepared to put her romantic life into hibernation simply because her husband had died. Of course she was utterly devastated, yet she also had her life to live. In 1910 she scandalised everyone by shedding her black carapace of mourning garments and appearing in a white gown adorned by a cheerful red rose. She was in love again. Her long-standing friendship with Pierre's old colleague Paul Langevin had become something more emotionally intense and sexual. In July that year they moved into an apartment conveniently near the Sorbonne and lived together as a couple. Unfortunately Langevin already had a wife and four children.

The second Nobel Prize only made life more difficult for Marie. She was infamous, in the sense that everyone had heard of her, and now all of Paris, and indeed much of the world, was eager to scrutinise her private affairs. French nationalists sneered at her Polish origins, anti-feminists were appalled that a woman had brought the confusions of the emotional life into the pure realms of science, and above all, she was an adulterous foreign harlot. People couldn't quite decide if she was evil incarnate or the world's most alluring scientific heroine. Marie wrote soberly to the French National Academy of Sciences, which had suggested that she postpone her 1911 trip to Stockholm until the dust settled. 'The prize has been awarded for the discovery of radium and polonium. I believe there is no connection between my scientific work and the facts of private life ... I cannot accept the idea that the value of scientific work should be influenced by libel and slander ... I am very saddened that you yourself are not of this opinion.' The Academy could only surrender to Marie's impeccable logic, and she was awarded her prize in the proper manner.

Marie Curie's death from leukaemia, on 4 July 1934, was almost certainly brought on by many years' prolonged exposure to radiation, although at 66 years of age, her life was not dramatically shorter than for most other women of her generation. Meantime, radium became popular as a quack medicine. It was also used for illuminating the dials of watches so that they could be read at night. Factory workers used fine-tipped paint brushes to apply the luminous paint, every once in a while licking the tips of the brushes to keep them in shape. No one had told them that radium might be dangerous. The so-called 'radium girls' suffered appallingly, first losing their teeth, then developing fatal cancers. A factory in New Jersey poisoned to death more than a hundred women, and the subsequent scandal contributed to major changes in the laws relating to factory safety, even if there was no helping the poor women in this case.

And no one knows how many patients the quack medics killed with their miracle radium cures. According to American Endocrine Laboratories, its *Radiendocrinator* was 'the last word in scientific manufacture'. It had to be pretty special to justify its $150 price tag. Presented in an embossed leatherette case, it was intended to be worn over the endocrine glands: the ovaries and testicles, thyroid and pancreas, 'which have so masterful a control over life and bodily health'. Men were advised to wear a *Radiendocrinator* under the scrotum at night. Radium was expensive for the manufacturers to source, so only the very *best* brands actually contained any radium. Most of the other devices and compounds on the market were fake. The victims of such cheapskate antics were the lucky ones.

The reluctant revolutionary

Pierre and Marie had been more than happy to discover new and extraordinary wonders of nature, despite the great costs to their health and comfort. A German contemporary, Max Planck, was not quite such a willing contributor to the atomic revolution. When he tried to unravel the secrets of radiation, he ended up deepening the mysteries.

Our historical impressions of Germany in the early 20th century are so badly skewed by the militaristic disasters of the First World War and the subsequent rise of Nazism that it's easy to forget certain German values that would have been prized in any country: doing the right thing by others, keeping one's word, protecting those less fortunate than oneself, and respecting the rule of law. Max Planck was the uprightest of upright citizens, and his faith in the benefits of a well-run society meshed perfectly well with his sense of duty towards individuals – and his deepest hopes for science, too. In the course of his long life he would be forced to question all his cherished assumptions, yet his morality would never fail him, even in the most frightening circumstances.

Planck was born in Kiel in the spring of 1858, and grew up in the era of German unification under Otto von Bismarck. Max's father, Julius, was a successful law professor. Everything about the Planck family, from its conservative Prussian values to its religion, spoke of orderliness and duty. At seventeen, Planck decided he might like to study physics, although he was warned by his teachers that all the worthwhile work in that subject had already been completed, so he would be dooming himself to a colourless career. Planck believed otherwise, and travelled to Berlin where he studied under two renowned physicists, Hermann Helmholtz, discoverer of the law of conservation of energy, and Gustav Kirchoff, an expert in electricity and thermal radiation. They may have been clever theorists but they were numbingly dull lecturers. Even so, Planck was inspired. Underlying all the blackboard scribblings and sub-divided specialisms there had to be, he felt, a single unifying set of laws. 'The outside world is something independent from man, something absolute, and the quest for the laws which apply to this absolute appeared to me as the most sublime scientific pursuit in life.'

In the mid-1890s Planck began to look at the problem of radiated heat. The classical view was that the wavelengths of energy given off by a hot object, whether as heat or light, must be infinitely adjustable from high to low frequency and all possible frequencies (or wavelengths) in between. After all, everything else in Newton's classical physics exhibited that sliding-scale smoothness. Light and heat waves were supposed to be smooth and sinuous, not stepped and jagged. Unfortunately, this belief led to an unacceptable conclusion. According to the laws known at that time, a hot object was supposed to give off energy at all possible frequencies, up to a certain maximum, depending on how hot it was. Since the number of 'in between' frequencies was supposedly unlimited, the total radiated energy should be infinite!

The main problem hinged on an idealised object called

the 'black body', which could emit or absorb electromagnetic radiation at all frequencies. In practice such a perfect thing does not exist, but in the mathematics prior to 1900, it *could* exist, and therefore some of its extreme outcomes had to be dealt with. It was known from laboratory tests that shorter wavelengths of electromagnetic energy were hotter than the longer wavelengths. Short-wave ultraviolet light packs much more of a punch than longer wavelength infra-red. To put it another way: the closer the waves are bunched together one behind the other, the higher the frequency of wave crests arriving at a certain point becomes, and the higher the heat energy climbs. So what was supposed to happen at the extreme ultraviolet end of the electromagnetic spectrum, where the wavelengths are very short indeed? Did the wavelengths eventually bunch together? If so, then what was to stop the waves bunching up so tight that the gaps between their crests vanished to nothing? The snag in the 'black body' concept was that zero wavelength had to deliver infinite energy ... This was known in the trade as the 'ultraviolet catastrophe'.

Obviously this is not what people were discovering in their laboratory experiments. Planck devised a mathematical fix for the problem, in which energy was absorbed or emitted in discrete bundles governed by a special number which he called 'the elementary quantum of action'. By mathematical trial and error he found the number to be 6.626×10^{-34}. This tiny yet *definite* value is known as h, or Planck's Constant. In December 1900 he announced this deceptively simple-looking relationship: energy = h x frequency. The ultraviolet catastrophe couldn't happen, because electromagnetic wavelengths could never shrink below h.

This discovery was not particularly pleasing to Planck, because he could not understand why his new number, h, had to have that particular value. He had invented a piece of mathematics that tamed the infinite range of possible wavelengths into discrete and finite chunks or packets, yet he had absolutely no physical model for how energy could be chopped up into these little pieces, other than to talk in a vague way about sub-microscopic 'vibrating oscillators' responsible for exchanging energy in specified bundles. He hoped that h would prove to be just a temporary solution until a better idea came along. It cannot be overstressed how unwelcome was this possibility that nature might have some underlying fragmentation or jerkiness at its heart. The instinct of a classical physicist was that all objects, distances, energies and forces could be subdivided ad infinitum, and were silky smooth in their transitions from large to small. The problem with the stepped transitions implied by h was that the gaps where certain energy values were *not* allowed to exist still had to be explained. And that mystery was beyond the science of 1900 to solve.

We would recognise the vibrating oscillator concept, today, as the electron shells of an atom, absorbing or emitting discrete photons of electromagnetic energy: radio, infra-red heat, visible light, ultraviolet, x-rays or gamma rays, depending on the energies involved. For Planck and his generation, atoms were still just statistical ideas, and the true connection (for instance) between x-rays and visible light – the fact that they were variations of the same pheno- menon – was far from clear. The wave nature of heat and light seemed unassailable, because laboratory experiments had proved time and again that they travelled as waves. Over the next decade and more, h became a useful tool in the mathematical armoury, yet no one, least of all Planck, believed it was real, because waves could not be subdivided, pellet-style, by his 'quantum of action'. He hoped that h would turn out to be just 'a fictional quantity, essentially an illusion representing only an empty play on formulas of no significance'. He worried that he might accidentally be taking physics into new and strange territory. Half a century later he recalled: 'My futile attempts to put the elementary quantum of action into the classical theory continued for a number of years and they cost me a great deal of effort … Here was something completely new, never heard of before, which seemed to require us to basically revise all our physical thinking.'

Planck's great achievement was to discover something that appalled his deepest instincts, yet never to shy away from it. Towards the end of his life he would be forced to abandon even more of his hopes.

The patent clerk

Five years later, in one of his great papers from 1905, the previously unknown patent clerk Albert Einstein suggested that light waves (and by implication, heat, radio, or any other electromagnetic waves) could be subdivided – and not just into abstract numbers, but actual particles. It's easy to think that he must surely have won his Nobel Prize for his world-famous theory of relativity. In fact he won the prize for his analysis of the 'photoelectric effect', a neat and subtle paper that was to cause far greater upsets in physics than relativity has ever done. Relativity brought a supremely satisfying and rounded sense of order to the universe at the grandest scales, while at the smallest scales, the photoelectric effect would eventually cause havoc in the scientific community.

In 1704 Newton had concluded that light 'is composed of tiny particles, or corpuscles, emitted by luminous bodies'. He also claimed that they must travel in straight lines, otherwise known as rays. The alternative wave-like properties of light were demonstrated by Thomas Young in the early 19th century. He shone light through two narrow slits, and saw, on a screen behind the slits, how a wave-like pattern of interference fringes was created. Where the two waves crossed over each other and crest met crest or trough met trough, the intensity was strengthened, revealing itself as a bright banding on the screen; and where trough met crest, each exactly cancelled the other, leaving darkness. The patterns made by Young were precisely similar to those made by waves in a glass water tank when they hit a barrier with two narrow openings. There seemed no doubt about it. Light propagated through space just like water waves. Young's contradiction of Newton's corpuscle theory offended the English scientific establishment, and was not therefore as immediately ground-breaking as it deserved to be. Meanwhile an experimenter across the Channel, Augustine Fresnel, came up with similar ideas, only to find that the French were equally opposed to thinking of light as a wave.

By 1900 no one doubted that light and heat travelled as waves. The question was one of medium rather than message. What did the waves travel through? The emptiness of outer space clearly did not prevent heat reaching us from the sun, nor light from the most distant stars. There had to be an invisible medium, a 'luminiferous aether' through which the waves propagated.

Einstein looked at experiments where light is shone onto a plate of metal, causing electricity to flow through the plate. The strength of this effect varies greatly depending on which metal is used, but zinc, for instance, is particularly good at generating a current (a flow of electrons) when exposed to light. The odd thing is that turning up the light, making it brighter, more intense, makes little difference to the strength of the electric current *if the light is the wrong colour*. On the

Figure 1. When waves intersect, a distinctive interference pattern emerges at the overlaps.

other hand, changing the colour of the light can produce dramatic results. Some colours will create a surge, others will have no effect. If you choose the right metal to work with, a powerful red lamp so bright it hurts the eyes will deliver no current, while a blue light so dim you can hardly see it will cause the current to flow. If light travels in waves – and remember, Young had proved back in 1801 that it does – then this should make no sense.

Einstein realised that the only way to explain the photoelectric effect was to say that, instead of being a wave as was generally accepted, light is actually made up of small packets of energy that behave like particles. He showed that red light cannot dislodge electrons because no individual red light particle, or 'photon',* has sufficient energy to shift an electron. It makes no difference turning up the light intensity, throwing yet more red photons in their countless trillions towards the zinc metal target. None of them will ever find their mark. But blue photons do have sufficient energy – and photons of ultraviolet light, which have even more energy, will give electrons such a kick that they fly off the metal altogether. It's not brightness but *frequency*, the spacing between waves per unit time, that signifies the energy carried by light. Blue light has a higher frequency than red light, and therefore blue light conveys more energy. And the frequencies, of course, were discretely stepped in accordance with Planck's new number, *h*.

In the wake of Einstein's paper, some of the more adventurous physicists began to talk about electromagnetic photons instead of heat or light waves. It was an idea that was too revolutionary for most classical tastes. After all, anyone could still perform the Young two-slit experiment and prove that light travelled in smooth waves, not condensed little bullets. Anyone could still reflect and focus

*Einstein never actually used the term 'photon' in his paper. As far as this author can tell, the first use of the word appears in a letter to the 18 December 1926 edition of *Nature* magazine from the American chemist Gilbert N. Lewis, in which he specifically claims credit for that term.

heat waves in a parabolic dish. Nothing had changed. Had it? Then again, perhaps it had. Einstein showed that particles of light and particles of matter interacted one-on-one. In his revolutionary paper, he wrote: 'When a light ray is spreading, the energy is not distributed continuously over ever-increasing spaces but consists of a finite number of energy quanta that are localised points in space, which move without dividing, and which can be absorbed or generated only as a whole.'

This fundamental carrier, the 'photon', is a particle of light that has no mass but can convey energy: a concept that is very tricky to reconcile with the notion of light as spread out waves propagating through space. All electromagnetic radiation is 'light', even though we can see only a limited range of it through our eyes. Photons are emitted by atoms when they are heated or energetically boosted in some way and then drop back down to their original state, shedding their excess energy in the form of photons. The more energy that is fed into the atoms to begin with, the higher the energy (frequency) of the photons emitted.

Visible light accounts for only a very thin sliver in the electromagnetic spectrum. Radio waves are also a form of light, but with longer wavelengths and lower energies. At the upper end of the light spectrum, x-rays and gamma rays have extremely short wavelengths; or to look at it another way, high frequencies, where the waves are packed very closely together. As Planck had discovered, high frequency means high energy. At this uppermost end of the spectrum, the waves are packed together extremely tightly, and the dividing line between energy and matter begins to blur ...

It was hard to imagine a wave compact enough to collide with a single electron, leaving all the other neighbouring electrons unaffected, but Einstein had shown that is what a photon can do. It was even stranger to think of photons smearing themselves through space to become waves the size of buildings, but that's what radio photons can do. None of it made intuitive sense, but the mathematical descriptions of these phenomena were among the most reliable tools of the human intellect, and the technologies based on them were about to be unleashed, from radio and the earliest telephones, to lightbulbs and hospital x-ray machines.

Planck had found that energy is lumpy, and Einstein had shown that electromagnetic waves consist of particles. Or was it the other way round – were particles waves? The confusions multiplied. And the next shock to the classical physicists was when the solid world of *matter* also turned out to be ambiguous. A noisy New Zealander found that the atom, that much-vaunted 'fundamental building block' of matter, consists mainly of nothing.

PART TWO

The Empty Atom

Our sense of the world used to be that solid things are solid. Everything is made of atoms, and in the first years of the 20th century atoms were also thought to be solid – perhaps like tiny billiard balls. In 1911 a gruff New Zealand-born experimenter called Ernest Rutherford made one of the most unsettling discoveries in history. The atom consists mainly of empty space.

Ernest Rutherford was born on 30 August 1871 in Nelson, New Zealand, the fourth child in a family of seven sons and five daughters. His father James Rutherford, a Scottish wheelwright, had emigrated to New Zealand in 1842. His mother, Martha Thompson, had been an English schoolteacher in her former life. She and her husband were keen that their children should benefit from a good education. In 1889 Ernest won a university scholarship and he proceeded to the University of New Zealand, Wellington, where he graduated four years later with a double first in Mathematics and Physical Science. In 1895 he was awarded another scholarship (or rather, he inherited one, when a rival turned down the same chance), this time to travel to England and go to Trinity College, Cambridge, as a research student at the Cavendish Laboratory under J.J. Thomson. The tall, gruff-spoken New Zealander farmboy discovered a world of arcane prejudices and class divisions. The Cavendish research students with their vacuum tubes and electro-mechanical experiments were looked down upon by the broader society of classical Cambridge academia. There were a good few snobbish dons who made no effort to distinguish, for instance, whether Rutherford was a New Zealander or an Australian.

There were benefits to compensate. At that time the Cavendish was one of the most respected laboratories in the world. Founded in 1874 with a grant from the Duke of Devonshire's family, its three directors since then had become household names. James Clerk Maxwell had unveiled the secrets of electromagnetic fields, Lord Rayleigh broke new ground in the study of light, and J.J. Thomson discovered the electron. Rutherford wrote home to his fiancée Mary Newton that he had struck up a cordial relationship with Thomson. With a scientist's characteristic knack for observing details, he recorded: 'a medium-sized man, very pleasant in conversation, and not fossilised at all. He shaves very badly and wears his hair rather long … he has a good head and has a couple of vertical furrows above his nose.' Rutherford spent three years with Thomson, first working on x-ray experiments, then moving towards investigations of his own, into the other equally mysterious uranium and radium rays recently discovered by Becquerel and the Curies. Rutherford's skill at designing elegant experiments from the simplest equipment was remarkable. He also had a real flair for knowing which investigations to make. A former colleague wrote of him: 'He could pay attention not so much to what Nature was saying as to what Nature was whispering. In this, Rutherford was an artist.'

In 1898 Rutherford made an important discovery simply by wrapping a sample of uranium in successive layers of aluminium foil. Three layers appeared to stabilise the radioactive emanations, but after that, adding another three layers made little difference. Yet more layers, and suddenly the radiation levels started to decline again. Rutherford decided that 'these experiments show that the uranium radiation is complex, and that there are present at least two distinct types of radiation – one that is very readily absorbed, which will be termed for convenience the alpha radiation, and the other of a more penetrative character, which will be termed the beta radiation.'

Henri Becquerel, the Curies and Rutherford between them proved that alpha particles were considerably less deflected in a magnetic field than beta particles. Just as they all suspected, the alphas were obviously more massive, and harder to divert from their flight paths. Just as significantly, the alphas were bent one way in the magnetic field, while the betas went the *other* way. Obviously they had opposite electrical charges! J.J. Thomson's beams of negatively charged corpuscles and Rutherford's beta radiation turned out to be exactly the same thing: electrons. On the other hand, x-rays could not be steered by a magnet. And nor could the final and most penetrating radiation they all encountered, which Rutherford called 'gamma' rays. In time it would become clearer that x-rays and gamma rays were not so much fragments of atoms, but photons of extremely high-energy light.

While Rutherford's scientific reputation prospered, his finances did not fare so well. He was anxious to find a proper source of employment so that he could marry Mary Newton. He also wanted a laboratory where he could continue to experiment with alpha rays without stumbling against Cambridge's arcane and pointless politics. McGill University, in Montreal, Canada, offered him everything he needed to fulfil both his scientific and personal ambitions. He might have been less successful without Madame Curie's occasional aid packages of pure uranium and radium to experiment with. He and Curie always maintained an extremely cordial relationship, although Rutherford's next experiment did cause Marie to wonder for a while if her New Zealander friend had strayed into territory that most scientists would have regarded as mythical nonsense. In 1902 Rutherford struck up a happy working relationship with Frederick Soddy, a young chemist at McGill. Together they discovered that a mildly radioactive metal called thorium emanated a gas, which then decayed into *solid* deposits on the walls of a glass container. Those deposits also decayed into something else, which they called 'thorium X'. Then came further radioactive decays into 'thorium A', 'thorium B' and 'thorium C'. The chemical traces were so tiny that they were impossible to identify with the equipment available, yet it seemed clear that the samples were transmuting from one substance to another in front of their eyes.

In a long tradition of pseudo-science, mysticism and esoteric ritual, the great alchemists of the Middle Ages searched in vain for the 'philosopher's stone', a legendary substance which was believed to be capable of transmuting everyday artisan's metals, such as copper and lead, into rare and expensive silver or gold. Even Isaac Newton held a torch for the more intellectual disciplines of alchemy, devoting as much time to this ancient craft as to his new mathematics. The quest for magical gold slowly transmuted into the painstaking and more accurate science known today as chemistry. By the turn of the 20th century no one but a few die-hard eccentrics believed any more in an alchemical world. Rutherford looked at the thorium experiments and wondered, bleakly, if he and Soddy might find themselves branded as alchemists. The permanence of the atom had never been in doubt before.

Rutherford and Soddy also identified the phenomenon of radioactive 'half life'. While their original thorium samples maintained a steady level of radioactivity for days or weeks on end, the radioactivity from some of the secondary decay products weakened dramatically after a few days. A precise pattern soon emerged. For any given radioactive sample, exactly half of its atoms would decay in a certain time. One of the thorium by-products, for instance, had a half-life of four days. The half-life of a sample of pitchblende uranium, on the other hand, turned out to be a great deal longer. Rutherford knew that most rocks contain traces of uranium, albeit in amounts far too small to be extractable, yet the radiation count gave him a clue about how *long* the uranium might have been locked inside a given rock. He calculated that pitchblende must be somewhere around 500 million years old: far older than many of his contemporaries supposed the age of the entire earth to be. Rutherford's initial results were tentative, but there could be no doubting that the earth had been here for long enough to support the slow, gradualistic evolution of animal and plant species which Charles Darwin had promoted in 1859. Rutherford had unseated the long-held biblical idea that the earth was no more than a few thousand years old.

Fifteen-inch shells

Rutherford spent nine years in Montreal until tempted back to England in 1907, when the University of Manchester invited him to direct its physics laboratory. He was pleased to find that the Mancunian industrialists and self-made businessfolk shared his distrust for snobberies and welcomed his practical flair for science. He also learned that one of his young assistants, a visiting German physicist named Hans Geiger, was just as keen as he was to make further investigations into the alpha particles. 'Geiger is a good man and works like a slave,' Rutherford wrote to a friend. Without him, 'I could never have found the time for the drudgery'. The booming-voiced New Zealander was always a little too eager for the next big thrill. He was usually at best when his energies were moderated by people with a steadier way of working. Brusque in his movements, he was not the most patient experimenter, even with his own most ingenious devices, and he often swore at the equipment – or his collaborators. Ernest Marsden, a young undergraduate assistant, was a protégé who knew not to take his boss's gruff manner too seriously. One day, just after Rutherford had checked on a fragile prism spectroscope and found it slightly misaligned, Marsden, working in another part of the darkened basement laboratory, felt a heavy hand on his shoulder. 'Did you move my prism?' Rutherford demanded to know, his voice quivering with rage.

'No,' Marsden replied calmly. Rutherford stomped off to find the culprit. Half an hour passed, and Marsden became aware of Rutherford's bulky frame re-entering the lab and sitting down next to him. 'Sorry,' he said, his voice much quieter now. These sporadic rages weren't taken seriously by any of his colleagues, except when his too-loud voice upset sensitive instruments. In fact his better moods were more dangerous than his bad ones. If a line of research was going badly, he'd mutter: 'Oh well. Onward Christian soldiers.' A more promising result would propel him into a full-throated table-shaking rendition. It seemed to be the only song he knew, or ever needed.

It was time, now, to see what alpha rays were made of. In 1908 Rutherford and Geiger enclosed a sample of radium inside a vacuum tube container with walls thin enough so that alpha rays could escape. Then they encased this whole assembly in another vacuum tube, but this time one with walls dense enough to block the rays. After running the experiment for several days, they passed an electric current through the space between the walls, where the alpha rays had essentially become trapped. When Rutherford examined the subsequent faint glow through the spectroscope, he found that it was characteristic of the light given off by helium. The simplicity of the experiment left no room for doubt. Alpha rays were streams of helium atoms. Tests with electromagnetic deflection had already showed that they had a positive charge. They could only be 'ions', helium atoms shredded of their electrons: plum puddings missing their plums.

These particles couldn't be seen in any conventional sense, of course. Each one was millions of times smaller than a flea. Yet even with the primitive equipment available in Edwardian England, they could be detected with ever more accuracy. A single particle striking a screen made of zinc sulphide creates a tiny flash of light, called a 'scintillation'. Rutherford and his team spent many hours in darkened laboratories, adjusting their eyes to the gloom until they could spot the flashes through a microscope lens. They had to be conscientious, positioning the screens in their narrow microscope tubes at varying angles to an experimental apparatus, front, back, sides, and many angles in between, all the while counting and logging their results. Rutherford was more than happy to delegate most of this repetitive work to his assistants. He did try it once, 'damned vigorously, and retired after two minutes'. Geiger was expert at the work, although he, too, would eventually be driven to creating the automatic scintillation counters that still bear his name.

Another valuable device had a charming and evocative name. It was called a 'cloud chamber'. In 1894 a young physicist, Charles Wilson, was standing at the summit of Ben Nevis, the highest mountain in Scotland. He was awed by the beauty of the sky, and noticed the strange prismatic coronas and halos that the sunlight seemed to cast in the mists and clouds. As soon as he was back in the lab, he built a cylindrical glass chamber the size of a large jam jar, and filled it with water vapour. Pulling a piston out of the device lowered the pressure until the vapour became very fine. Wilson's original intention was to shine light through the vapour and recreate, in miniature, those prismatic halos he had seen from Ben Nevis. Somewhat to his puzzlement, he couldn't entirely prevent visible droplets of water condensing out of his miniature mist, apparently ruining the optically smooth effect he was after. His equipment was extremely clean and dust-free. What, then, was causing those occasional droplets to form, as though they had found some impurities in the vapour to latch on to and condense around?

By 1911 he had found the answer by firing the newly discovered alpha and beta particles through his chamber. He surmised, correctly, that the particles were knocking negative charge (dislodging electrons) from the water molecules, creating tiny islands of positive charge, which in turn attracted the negative charges of other nearby water molecules. Droplets condensed around these tiny islands of perturbation. With a conventional optical microscope, the vapour trails left in an alpha particle's wake could clearly be seen.

That same year, Rutherford published the results of an experiment in which he fired a stream of alpha particles through a single very thin gold foil, hoping to discover more about the internal structure of atoms. His intention was to measure how the positively charged alpha particles might be deflected or 'scattered' on their way through the foil by the electrical charges of gold atoms; but since he'd already observed alpha particles shooting through several layers of aluminium foil, he expected them to fly through this latest obstacle with similar ease, albeit slightly diverted in their courses.

It was Geiger who suggested that the young undergraduate Marsden be given the necessary but presumably unexciting task of checking that none of the alpha particles were scattered to unusually extreme angles – say, back towards where they'd come from. And yes, most of the alpha particles did penetrate as easily as bullets punching through cobwebs, but Marsden found that *some* of them rebounded off the foil as though they had slammed into an unbreachable wall. After checking and rechecking his results, he took them to his astonished boss. 'It was quite the most incredible event that has ever happened to me in my life,' Rutherford recalled. 'It was as though you had fired a fifteen-inch shell at a piece of tissue paper and it had come back and hit you.'

Thomson's spongy plum pudding model of the atom had to be wrong. Only a very strong and densely packed cluster of positive charge could have repelled the energetic alpha particles. The fact that most of the alpha particles passed through the foil, while only a very small fraction bounced back, told Rutherford something amazing about the atom.

It consisted almost entirely of empty space. Only very occasionally did an alpha particle encounter one of those islands of positive charge head-on; and only then did it bounce back towards the startled scientists in Rutherford's lab. A new model of the atom was born that year: one in which the positive charge of an atom, and almost all of its mass, is concentrated in a 'nucleus', surrounded by a cloud of electrons. When Rutherford contemplated the size of the nucleus in comparison to the overall reach of an atom's electrical and physical influences, he said it was like 'a gnat in the Albert Hall'. This may be a visually compelling image, yet it doesn't convey to us that the little gnat is thousands of times heavier than the building it inhabits.

We still use this approximate Rutherford-era model of the atom today, because it's good enough to explain the basic chemistry that we learn in school. A matching amount of positive charge inside a nucleus precisely balances out the negatively charged electrons. If an electron is knocked away from the cloud, or another one added, then the atom becomes an electrically imbalanced 'ion'. This was a fantastically important discovery. Ions lie at the heart of almost all of chemistry, because the imbalance of charges sets up a tremendous tension that can be resolved only by some kind of reaction that re-establishes the atom's neutral charge.

Yet Rutherford's atom posed new and extremely puzzling questions. If opposite charges attracted each other, why weren't the negative electrons swirling around the atom dragged towards the positively charged nucleus? Rutherford decided that the inwards pull of the charges must be balanced by the outwards-acting centrifugal force of the flying electrons. Unfortunately, if they were hurtling round and round the nucleus at the necessary speed, then they should have been obeying Maxwell's laws relating to electricity. Basically, when an electron moves, it constitutes a flow of electricity; and just like any flowing current, it gives off energy as a surrounding electromagnetic field. Whatever produces energy must eventually dissipate, for no energy source can last forever. In fact an electron should exhaust itself and spiral down towards the atomic nucleus in a few billionths of a second … How come electrons can remain outside the nucleus at all? Not just for whole seconds, hours and minutes, but potentially for ever?

The 'Great Dane'

In the same year that Rutherford and Marsden discovered the nucleus, a 26-year-old Danish scientist, Niels Bohr, was struggling to win J.J. Thomson's attention at the Cavendish, and somehow he just wasn't getting through. He found a more receptive listener and a better working atmosphere in Manchester, where Rutherford found him to be 'quite the most intelligent young man I ever met'. Bohr seized his chance to prove what he already suspected – that a new kind of physics would be required to investigate the atom: a physics that abandoned more than two centuries' worth of Newtonian assumptions about how the world works. In an April 1913 paper entitled 'On the Constitution of Atoms and Molecules', produced with Rutherford's support, Bohr warned about 'the inadequacies of the classical electro-dynamics in describing the behaviour of systems of atomic size'.

Bohr's mannerisms and styles of speech were so con-voluted, and his voice so soft, that it was often quite difficult for his listeners to follow his arguments. Even the shy and socially gauche English physicist Paul Dirac (more about him later) said to Bohr once: 'Were you never taught in school that before you begin a sentence you should have some plan as to how you are going to finish it?' Yet those who had the patience to hear him out were invariably inspired by his intelligence, his personal warmth, and by the profound depth of his thinking. He was slow in his deliberations, even about the most apparently trivial things. Friends would take him to the cinema, for instance, and he'd whisper to them in the dark, apparently stymied by the most obvious plot points. 'Is this the sister of the cowboy who tried to steal the herd of cattle belonging to the brother-in-law?' Movies aside, once he understood something, he understood it better than any other physicist of his generation. He was also obstinate and strong-minded, and perhaps it was wise for his colleagues to keep in mind his youthful passion for football. In 1908 he was reserve goalkeeper for Denmark's Olympic team (his much-loved brother Harald, a highly respected mathematician in his own right, actually made the Olympic squad). Sometimes Bohr's attitude towards a favoured scientific idea was as dogged as a keeper's guardianship of his goalposts.

When it came to the problem of electrons spiralling into the nucleus in fractions of a second, Bohr's solution was to make up a new rule, while ignoring for the moment how it might actually work. He suggested that each electron occupies a stable orbit around the nucleus, and as long as it remains undisturbed it simply doesn't radiate any energy. On the other hand, it can 'jump' down from one orbital level to another, emitting a discrete packet of energy as it does so. Correspondingly, it can absorb a packet of energy from the outside world and jump 'up' to a higher orbit. Everything happens in discrete steps. There are certain permitted orbits, and certain intensities of energy packet that can be given off or absorbed, but no permitted states in between. Everything depends, too, on those packets of energy being exchanged between the atom and the outside world. Leave it alone, and its electrons will calmly remain in their orbits. Bohr argued mathematically that there was a close relationship between his notion of steps and jumps and the 'elementary quantum of action' described by Max Planck in 1900. Bohr was not surprised to find that Planck's constant, h, was perfect for calculating the discrete energy levels needed to make the orbital jumps.

The spectra of light emitted by elements heated to different temperatures now made a new and startling kind of sense. Einstein's photoelectric paper showed that a photon of light can knock an electron free of its host atom. Bohr showed how a photon with a slightly lower energy can kick an electron into a higher orbit without actually dislodging it. When the electron falls back down to a lower orbit, it sheds its excess energy by *emitting* a new (but lower-energy) photon, which we can see, or register in our instruments. Bohr showed that the further an electron falls back, the higher the energy of the photon it releases. Think of an iron bar heated in a furnace until it glows red. The electrons in the iron atoms are being kicked a few orbits up, and when they fall back they give off low-energy red light. Now heat the bar more ferociously until it glows blue-white. The electrons are boosted to yet higher orbits, and when they fall back down, they fall even further than before. Bohr worked out that the blue-white light that we see as the iron bar glows ever brighter is a consequence of the higher-energy photons of light streaming away from it. He had found a neat and exact connection between the colour of light given off by heated elements and the energy levels of the electrons in the atoms.

This theory wasn't any kind of an explanation for how the atom worked, let alone what it really was. He knew he was investigating abstract ideas which very few ordinary people, and perhaps few scientists either, could grasp intuitively. He was starting to believe that 'there can be no descriptive account of the structure of the atom. All such accounts must necessarily be based on classical concepts which no longer apply. We lack a language in which we can make ourselves understood.' It was a key moment in the history of science. Until this moment, scientists had expected to be able to understand nature using a language that made sense to the human mind. Bohr threatened to send the atom, the very bedrock of material existence, into a realm beyond our mental grasp. It was a disturbing idea whose shockwaves are still being felt to this day. Where Bohr began tentatively to tread, others would soon follow. The tumult had barely begun.

Bohr also said something else of profound significance. The moment *when* an electron happened to jump from one orbit to another was not predetermined. It was a matter of chance, of probability. The implications of this idea did not immediately transform the world of science, but in the coming decade this notion of probability would revolutionise everything we thought we knew about cause and effect. It would cause us to question the very nature of reality.

For now, the main problem in the Bohr 'solar system' model of the atom was that his calculations worked in detail only for the behaviour of a hydrogen atom with a single electron orbiting around it. But all the other atoms in the periodic table had multiple electrons: swarms of the wretched things. Helium, the next element in the chain, contained two electrons. Then came lithium with three, beryllium with four, boron with five, carbon with six, nitrogen with seven, oxygen with eight, and so on. And these were just the light elements. At the heavy end of the periodic table there were atoms with dozens of electrons that had to be slotted into their proper orbits.

Bohr returned to Copenhagen quite the rising star, yet the academic posts offered to him seemed too unimaginative after his breakthroughs at Manchester. On 10 March 1914 he wrote to the Danish Department of Educational Affairs. 'The undersigned takes the liberty of petitioning the department to bring about the founding of a professorship in theoretical physics at the university [of Copenhagen] and in addition to possibly entrust me with that position.' Nothing could happen overnight. Bohr returned to Manchester while the 1914–18 war played out its terrible course. Denmark remained neutral throughout. At war's end, having offended no one, it was in an excellent position to serve as unbiased mediator amid the wreckage of international science.

Bohr's decent manner only added to his appeal. Public and commercial organisations alike were happy to hear his plans. In 1921 he inaugurated a purpose-built three-storey building, the Institute for Theoretical Physics in Copenhagen, funded by the Carlsberg brewing company in partnership with the Danish government. Bohr, his wife Margarethe and their two boys lived on the second floor, while residential guests occupied the third. The ground floor was given over to lecture rooms, libraries and offices, and the basement held laboratories for practical experimentation. It was a happy scientific paradise, in many ways an innocent place, with a warm atmosphere and a lively social buzz. Scientists from around the world were drawn to Copenhagen and the court of the 'Great Dane'. Even across the Atlantic, the warm glow of excitement given off by the place was almost palpable. When space began to run short, a generous $40,000 (equivalent to nearly half a million in today's dollars) was granted by the Rockefeller Foundation in America to expand the building.

Other men in Bohr's position might have lorded it. His rule over the empire of early atomic physics was benign, and his personal needs were modest. The debates that raged in his Institute in the late 1920s caused a revolution in science. Bohr's passions were just as great as anyone else's, and he could be relentless when challenging someone's ideas, yet the arguments almost never descended into personal animosity. If there has ever been a time of pure and joyous perfection in the story of humankind's relationship with the atom, then it was probably in Copenhagen during the 1920s, where an optimistic and passionate group of physicists argued by day and caroused by night, fomenting a revolution in science privileged to very few people outside their tight-knit circle. They disagreed loudly and sometimes bitterly about the meanings of their theories, but all of them adored the kinship and common ownership of this great adventure. The atom brought them together as close friends. And all too soon, it would divide them again.

PART THREE

Not Even Wrong

In the 1920s a brilliant and argumentative coterie of European physicists generated some absolutely bizarre theories of the atom that made no sense to Albert Einstein, the world's most respected scientist. He fought a last-ditch battle in defence of sanity. And he lost.

The Apollo 11 astronaut Michael Collins was once asked: what special qualification did an astronaut need to join one of NASA's historic flights to the moon? He replied: 'Above all, you need to have been born at the right time.' He then demonstrated that he and most of his colleagues had been born within plus or minus two or three years of 1930, and so were just the right age to participate when the rocket era came along. Something similar applied to the theoretical physicists who reshaped science between the years 1920 and 1925. To be a young and scientifically inclined man in the first two decades of the new century, you could not help but be captivated by unparalleled excitements: the Curies and their radiation experiments, Planck's quantum of action, and of course, Einstein's relativity. The historian Thomas Kuhn has given us a phrase for such a moment of upheaval in science: the 'paradigm shift', in which the old ways of understanding the world are unravelled and a completely different mental landscape is born. In the 1920s that new landscape came to be known as 'quantum theory', or more properly, 'quantum mechanics'.

As a student in Munich, Wolfgang Pauli published three

papers on relativity, and then his professor, Arnold Sommerfeld, asked him to write an encyclopaedia article on the same theme, treating the young man as a completely reliable source. Einstein checked the text and wrote a glowing review. 'Whoever studies this mature and grandly conceived work might not believe that its author is a twenty-one year old man. One wonders what to admire most … the complete treatment of the subject matter, or the sureness of critical appraisal.'

Following his graduation in Munich, Pauli was appointed to Göttingen in October 1921 as Max Born's assistant. Here he first met Niels Bohr, who immediately invited him to Copenhagen, where the new Institute was just opening for business. Pauli spent a year there before moving back to the University in Hamburg to become a lecturer. Like so many other physicists, he travelled restlessly around Europe, attending seminars and meeting the other major figures in his trade. His most frequent port of call was always the Institute in Copenhagen. He and Bohr struck up an intense and peculiar friendship, for one of Pauli's greatest talents was to be a scathing critic of careless ideas, yet somehow without actually causing anyone to dislike him. Bohr loved a meaty argument, and gave as good as he got, albeit without Pauli's knack for swift savagery. Pauli would shout: 'Shut up, you are being an idiot. I will not listen to another word!' Then he would try to reawaken the fighting spirit in his victim. 'Your job is, every time I say something, contradict me with the strongest arguments.'

Pauli's put-downs became legendary. On one occasion he returned a young colleague's paper with a spectacularly dismissive comment. 'This is not even wrong.' He told someone else that he didn't mind if they thought slowly, but he wished they wouldn't publish faster than they could think. He attended a lecture by Einstein, where even his respect for the great man was tinged by his acidic wit. Someone in the audience interrupted with a question, and Pauli called back: 'You know, some of Einstein's ideas are not so stupid.' Even when Pauli wasn't at the Institute, Bohr would still spar with him aloud, rehearsing his arguments, and always asking colleagues: 'Yes, but what would Pauli think?' In a wandering court of European theoretical princes, Pauli was the balloon-bursting jester who was allowed to speak whatever was not supposed to be spoken.

Werner Heisenberg, a contemporary from student days in Munich, never took offence. 'He was extremely critical. I don't know how many times he told me, "You are a complete fool", and so on. That helped a lot.' He once told Heisenberg: 'Only someone who lacks a thorough understanding of classical physics could think like you, so you have an advantage there. But remember, ignorance is no guarantee of success.' When Heisenberg expressed confidence with a theory he was working on, he unwisely remarked to Pauli that 'only the technical details are missing'. Pauli scrawled a jagged rectangle on a scrap of paper. 'This is to show I can paint like Titian. Only the technical details are missing.'

Experimenters learned to be wary of him, for he was a clumsy man with a roly-poly figure, quite incapable of using the simplest piece of machinery without mishap. Bohr delighted in ascribing every experimental disaster at the Copenhagen Institute to Pauli. He had only to walk into a room and some piece of equipment would break down. When one day an apparatus exploded for no apparent reason at the physics lab in Göttingen, Pauli was nowhere to be seen. The proper order of the universe was restored when enquiries showed that a train carrying Pauli from Zürich to Copenhagen had stopped briefly at Göttingen station at exactly the same time as the explosion occurred. That was the 'Pauli effect'.

Of course Pauli was his own worst critic, and this prevented him, perhaps, from making the great intuitive leaps into the dark that some of his contemporaries achieved; Heisenberg, for instance, as we'll soon see. Pauli hated any detail to be incorrect, or any argument to be less than watertight. He lacked an essential ability for any theorist:

the courage to hazard bold new ideas and risk being *wrong*. Even so, his role as arch-critic was deeply valued, for physics at its best is a craft that welcomes the most intense debate. His fast wit was entertaining, his acidic scorn left his friends and colleagues largely unscathed, and his technical criticisms they all took very seriously. And Pauli did deliver one truly brilliant piece of work. It was called the 'exclusion principle'.

Bohr had thought long and hard about the problem of atoms more complicated than hydrogen, with its single electron. How were the six electrons in a carbon atom, for instance, or the seven electrons in nitrogen and the eight in oxygen supposed to arrange themselves in his orbital model? He drew a number of schemes, with some electrons occupying perfectly circular orbits, and others following steeply elliptical paths. Some of his sketches looked like complicated arrays of flower petals, and none of them satisfied the quest for a unified rule about how electrons co-existed in an atom.

Pauli worked with the set of 'quantum' numbers that described an electron's orbital level and its possible energies or 'states'. There were three numbers corresponding, in simple terms, to an electron's freedom of movement in three dimensions of space. Pauli added a fourth number, associated with a quality called 'spin'. This is a difficult concept, so we'll look at a visual analogy. If two fast-spinning balls collide, then the spin swapped between them on contact is an important factor, as well as just the two different directions that the balls came from originally, and how fast they were moving when they collided. Snooker parlours and pool halls are haunted by players who can transmit spin from one ball to another with great skill. There are two kinds of spin: clockwise and anti-clockwise. Now imagine spin as a quality that electrons possess, but discard any simplistic image of them actually rotating … Anyway, Pauli decided that spin should be included as a fourth quantum number. In 1925 he delivered his famous ruling. No electrons with the same four quantum numbers can occupy the same orbit. If two electrons do share an orbit, then they have to have opposite spin numbers. Three electrons in the same orbit is impossible, because there are only two kinds of spin to share out between them, so there would be a clash of similar quantum numbers. Pauli had explained why electrons don't simply fall down into the nucleus. They are simply not all *allowed* to occupy that same space.

Rutherford had found the atom to consist almost entirely of empty space. Bohr had found a rule, albeit slightly arbitrary, preventing the electron of a hydrogen atom from spiralling down towards the nucleus. Now Pauli had worked out how to keep swarms of multiple electrons in their proper places. This was no small achievement. It explained how the solidity of all the matter in the universe is prevented from vanishing into nothingness when an atom is at rest. Quite apart from his contribution to physics, Pauli had suddenly made the rules of chemistry much easier to understand, for the bonds between atoms are determined by the interactions of their outermost electron levels, or 'shells'.

As far back as 1869, the Russian chemist Dmitri Mendeleev had worked out that certain properties of the elements, such as their mass per unit volume, and their chemical reactivity, had a discernible pattern. He created a table that listed the atomic weights of the elements in rows, and their shared chemical characteristics in columns. Blank spaces in the table were assumed to belong to unknown elements. It was a brilliant piece of work, undertaken by a man living in an age when the atom was still the wispiest abstraction. By the mid-20th century the blank slots in Mendeleev's table were fast being filled in, and Pauli's electron shells gave an explanation for all the chemistry. Take gold and mercury, for instance. A gold atom has 79 electrons, while mercury has 80. Gold is solid at room temperature, and chemically inert. It won't react with any other element. By contrast, mercury is liquid at room temperature, and highly chemically reactive. It takes a difference of only one electron to give gold and mercury their very different chemical characteristics. The jostling of electrons as they fill their allowable slots in the electron shells makes all the difference.

Not that Pauli had any love of chemistry, or chemists. Near the end of his appointment in Hamburg, he suffered a series of catastrophes. The first was the suicide of his mother in 1927, after she discovered that his father had been having an affair. The second disaster was that his father then married the other woman, the 'evil stepmother' as Pauli called her. Shortly after moving to Zürich in an attempt to clear his mind, Pauli married for the first time, a dancing girl called Käthe Margarethe Deppner. The relationship was a disaster, lasting less than a year. Käthe's greatest betrayal was in her choice of lover to run away with. 'Had she taken a bullfighter I would have understood. But an ordinary chemist!'

Pauli now began a long and close friendship with the pioneering psychologist Carl Jung. Many scientists are distrustful of psychology or any other field of thought where the terms are vague and the technical details unspecified. Pauli found some comfort from his distress when Jung convinced him that his argumentative nature was a reflection of his self-hatred. His relationships with women were coloured by grief on his mother's behalf and his anger towards his father's lover. This all might seem rather obvious to us today, from the simplest reading of Pauli's biography. Ours is an age that takes many forms of psychoanalysis pretty much for granted. For Pauli it was a revelation. He became fascinated by the apparent gulf between psychology and science, and would later write: 'It is my personal opinion that in the science of the future, reality will neither be "psychic" nor "physical", but somehow both and somehow neither.' He was among the first modern physicists to worry deeply about the interplay between the material world outside of us and the mental universe within us. Niels Bohr thought about these problems too. Throughout his life one of his favourite sayings, taken from an old philosophy book he'd encountered as a student, was this: 'We are actors as well as observers in the drama.' Waiting in the wings was a man who thought he could *prove* the truth of that sentiment.

More trouble with waves

Pauli's exclusion calculations worked beautifully as mathematical rules, yet the forbidden gaps between allowable electron orbits still needed a proper physical explanation. When the answers began to arrive, they didn't look like answers at all. They seemed more like paradoxes too profound for the human imagination to grasp. The first traditional idea that had to be abandoned was that electrons were point-like particles, reliable little nuggets or 'building blocks' of matter. The closer you looked, the more nebulous they became.

Louis de Broglie was a French aristocrat so wealthy that he could afford to build his own laboratory just off the Champs Elysées, where he dedicated himself to his one true love, physics. And what fascinated him was this. How come every time any physicist talked about a quantum of energy – or a photon, for that matter – they couldn't help talking about *frequencies*? How could a single bullet-like packet of energy have such a thing as frequency?

In 1923 de Broglie argued that if photons could be described with wave mathematics, then so could the electron. The strange business of electrons jumping from one orbit to another without occupying any intermediate spaces was not so puzzling if they were shimmering along on wave-like paths. There's no such thing as a fraction of a wave, he reasoned. Only whole crests and whole troughs exist, so there could be no vague 'in-between' orbits for an electron to occupy. The vacant gaps between the orbits are simply regions where crests overlap with troughs and cancel each other out. If matter had a wave-like component *and* a particle component, then Bohr's atom could make sense. It was a brilliant and important idea, but it still didn't get to the heart of the problem. Which is the electron, and which the wave? Did one travel upon the other, or were they one and the same thing? How could one visualise the atom as a physical entity, instead of just an abstract maze of conflicting mathematical descriptions?

One man insisted that there was nothing to visualise. When it came to the atom, there *were* no real particles, no real waves, and no hard and fast realities. The only thing he was sure about was that everything was uncertain.

Werner Heisenberg

Werner Heisenberg certainly was not born into a tradition of revolutionary thinking. His father, August, was an authoritarian schoolmaster, stiff, tightly controlling and precise. Werner's mother Annie lived, so it seemed, entirely to fulfil her husband's needs. The Heisenberg household in Würzburg expressed all the most rigid and self-disciplined aspects of the Prussio-German Empire at the turn of the 20th century. In 1910 the family relocated to Munich, where August had been appointed to the prestigious Chair of Greek Philology at the University. He introduced Werner to the Greek scientist-philosophers, and encouraged his talented musicianship at the piano; but the boy was not alone in seeking his father's approval. He had a twin brother, Erwin. Their psychologically unfortunate battle for favour, stoked by August, culminated in a ferocious battle with wooden chairs when they were in their early teens. For the rest of their lives the twins had as little as possible to do with each other.

The Great War of 1914–18 destroyed the certainties and illusions by which the Heisenberg family had lived. Bourgeois respectability was at an end, and the militaristic idealism of Empire had lost its steely sheen. The old guard had betrayed the younger generation, leaving a shattered

Germany for their inheritance. Eighteen-year-old Werner took a stance that was both rebellious and backwards-looking. He became the leader of a youth movement called the Deutscher Neupfadfinder, the 'German New Boy Scouts'. They hiked and camped in the hills, yearning for an innocent return to nature and a simpler and less morally ruined way of life. With its flawed romanticism and giddy hopes for a better and less squalid Germany, the Neupfadfinder was just the kind of gang of child-idealists that would one day become absorbed into the darker manipulations of the Nazi Party, although for now, Jewish boys were readily accepted into the troupe. Heisenberg was never an anti-Semite.

In 1920 Heisenberg entered the University of Munich to study mathematics, but he and his professor didn't get along, so he transferred to physics, and was lucky enough to attend a lecture given by Niels Bohr at Göttingen. He was electrified, and Bohr's model of the atom fascinated him the most. Bohr took him aside afterwards, and they arranged to go on a long walk in the country. It was the beginning of a great friendship: one that was destined not to last.

Heisenberg's first academic position was at the University of Göttingen, where he became an assistant to Max Born. His new mentor was charmed by this fit-looking young man 'with short fair hair, clear bright eyes, and a charming expression'. It took a few years for his greatest moment of inspiration to strike, but in 1925 the time arrived. He fell ill with hay fever, and Born granted him a fortnight's leave, which he spent on a windswept and clear-aired little North Sea island called Helgoland. As he stared out to sea and recovered his spirits, it suddenly hit him that the quest to reconcile waves and particles was not just fruitless, it was actually getting in the way. At Göttingen he had absorbed a 'positivist' view of science. Only facts and data that could be positively verified had any meaning. The only truths that mattered were those that could be observed and measured in the lab. There was no point in talking about fanciful and *un*measurable qualities of an atom, such as what it might 'really' be like. Similarly, the orbit of an electron could not be observed, and therefore had no reality. The entire notion was just a loose way of talking, a circle scribbled on a piece of paper. Ordinary language generated misleading models in the scientific imagination, Heisenberg realised. As he told an interviewer many years later, he became convinced that 'the words we use to describe experiments have only a limited range of applicability. That is a fundamental paradox which we have to confront. We cannot avoid it. We have simply to cope with it.' Bohr had already warned that 'there can be no descriptive account of the structure of the atom.' One of the most unsettling ideas in all of science had found a powerful pair of champions.

In similar vein, Heisenberg decided that the 'paths' supposedly taken by particles as they whizzed through an experiment were also nothing more than linguistic fictions invented after the fact. No one ever *saw* a particle traversing a particular path. Thomson never actually saw his neat, narrow beam of electrons. What he saw was a scintillation on a screen, from which he inferred the existence of the beam. Likewise, Rutherford never saw an alpha particle shooting from one place to another. All he had to go on was the subsequent impacts on a screen. The journey in between a radioactive source and a detecting instrument could be worked out only in retrospect, and the mathematical descriptions always led to unresolvable contradictions. Some paths taken by particles seemed as if they *must have been* wave-like, especially when viewed with instruments that measured wave diffraction effects; and some wave progressions seemed as if they *must have been* line-like paths, especially when traversing a cloud chamber. But what if the paths and waves didn't actually exist, and the only reliable piece of knowledge one could obtain was the measurable outcome of a particle's journey, and nothing more? If you looked for a wave, Heisenberg argued, you'd find the pattern such as a wave might leave on a screen. If you looked for a particle, then you'd find a track such as a particle might leave in a cloud of vapour. You would never find an actual wave or an actual particle.

He saw, too, that when he multiplied certain elements in the descriptions of atomic behaviour, it made a difference which way around the multiplications were made. To put it another way: ordinarily, 2 × 3 should give the same answer as 3 × 2. It's 6, whichever way around one plays the numbers. In conjunction with Max Born, his professor at Göttingen, Heisenberg created 'matrix mechanics', in which certain terms to be multiplied were arranged in huge brackets and then mapped onto numbers in equivalent positions in other sets of brackets, so that each term would be applied only to a specific partner term in the other matrix. This complicated procedure was abstract in the extreme. It couldn't be *visualised* in the mind's eye.

And the fact that the calculations worked only in a specific order carried a shocking freight. All classical physics depends on the interplay of masses, forces and energies being so predictable that the calculations can apply just as well backwards as forwards. If you roll one ball into another one, and the second ball moves a couple of inches, you'll find that 'playing the movie in reverse' still makes a realistic and completely believable story, where this time the second ball rolls a couple of inches, and the first ball is nudged. The mathematics is exactly the same in both cases. Heisenberg's matrix mechanics destroyed that comforting idea. It implied that an electron's behaviour is not fixed in advance. Instead it's a shimmering cloud of options and likelihoods and possibilities. There is never anything more than a *probability* that an electron might behave in a particular way, or land at a particular point on a detector. Matrix mechanics was essentially a betting system. There were terms analysing the possible outcomes of an electron's history, with some behaviours being extremely unlikely, and others being a little more likely, and some very likely, but *none* of those behaviours could be predicted in advance with any certainty.

This also meant that you couldn't work backwards from your experimental results to infer the starting conditions. Heisenberg's matrix-mechanical movie was impossible to play in reverse: impossible, even, to view as a 'motion picture' of any kind.

Waves of emotion

Whenever Albert Einstein lost his way in physics, he often turned to some affair of the heart to revive his spirits, even if those diversions caused ructions in his family life. Brains can be attractive when matched with some degree of personal charm, although the pleasing aspects of great physicists can so often be accompanied by self-centredness. Einstein once remarked that it was unreasonable for a man to be forced into a lifetime of marriage on the basis of a fleeting passion back in his youthful days. If the world of atomic physics can be said to have had any truly sexy superstars, then the mesmerically appealing Richard Feynman (more about him later) has to count as one of the most attractive men in the field; but while his many libidinous liaisons around the world only added to his allure, Erwin Schrödinger was the braver man, for he was an adulterer and enthusiastic womaniser in an age when louche behaviour could ruin your career.

Schrödinger was born in Vienna in 1887. His father Rudolf ran a small linoleum factory, which he had inherited from his own father, but his real loves were botany and Italian painting. His mother, Emily Bauer, was half English. Her side of the family had its roots in Leamington Spa. Even as a child, Schrödinger learnt English as easily as German. He didn't go to elementary school, but took lessons at home from a private tutor up to the age of ten, and he didn't enter the traditional 'Gymnasium' school system until autumn 1898. At eleven years old he was quite a late school starter, especially as he'd just enjoyed a long holiday in England that cut into his first term. It didn't make much difference. A schoolmate recalled the ease with which he kept up with all his lessons. 'It was possible for our professor to call Schrödinger immediately to the blackboard and to set him problems, which he solved with playful facility.'

Schrödinger graduated from the Akademisches Gymnasium in 1906 and went to the University of Vienna. In theoretical physics he studied analytical mechanics: basically applications of differential equations, the mathematics used for manipulating constantly shifting values. For the non-mathematically-minded, these are pretty daunting tools, but it's not too hard to understand how they are used. It's simple to calculate, for instance, the total number of bricks required to build a house of a given size, because the values on either side of the relevant equation are fixed, even if you don't know all of them until you've crunched the numbers. The world seldom stays that still, and physical entities are not usually neat and square, like a house or a brick. Oval pebbles roll down hills, water flows out of spherical tanks into cylindrical ones. Gears of different sizes mesh together and turn, constantly changing their overall state (you may well have a differential gearbox in your car). Planets in their orbits interfere with each other's gravity fields in a restless, ever-shifting interplay. Differential equations explore the properties of dynamic systems. This bias to Schrödinger's education would one day fuel a furious debate with Heisenberg. What kind of mathematics was best for describing the behaviour of an atom?

On 20 May 1910, Schrödinger was awarded his doctorate for a dissertation on the effect of damp air on electrical conductors and insulators. After this he undertook voluntary military service. Then he was appointed to an assistantship at Vienna, where he had a chance to do some serious hands-on experimental work instead of merely playing around with mathematical abstractions. In 1914 he was in uniform once again. He experienced combat and was awarded a citation for his command of a gun battery. His life thereafter was restless and peripatetic, as seems to have been the case with so many of the 20th century's great physicists, and especially those driven here and there by wars and politics as well as by the more conventional drives of academic advancement. He was appointed a professor in Stuttgart, then Breslau, then Zürich, where he settled for six of his happiest years and completed his best and most important work. In 1927 he was appointed as successor to the great Max Planck as head of physics at the Kaiser Wilhelm Institute in Berlin. When Hitler seized absolute power in 1933, Schrödinger became yet another of the great flood of disenchanted or frightened people to escape Germany in search of freedom. He was a Catholic Austrian who could have made a career for himself in the emerging German Reich, except that the Nazi ethos and its anti-Semitism disgusted him. He gladly accepted a fellowship at Magdalen College, Oxford.

It was a close-run thing. He'd attracted a job offer from Princeton, but neither the Americans nor the stuffed shirts at Oxford were happy about a man who planned to arrive with his wife Annemarie Bertel *and* his mistress Hilde, pregnant with his first child. Oxford eventually made a pragmatic decision in his favour. The unconventional domestic arrangements of a foreigner did not necessarily pose any risk to the accepted marital conventions of England. Annemarie seemed to have been even more understanding. She said once that 'life with a race horse is not as easy as life with a caged canary, but I'd rather be with a race horse.' The sexual life was almost as important to Schrödinger as his work in physics. He seemed always to be chasing beauty, whether in the form of a woman or a scientific truth. 'I put beauty before science,' he wrote to Max Born. 'We are always longing for our neighbour's housewife and for the perfection we are least likely to achieve.' This restlessness expressed itself geographically too. In 1936 Schrödinger went back to Austria, although by now the Nazis considered him a traitor. He packed his bags once more and headed to Rome, then Dublin, where the Irish Prime Minister Eamon de Valera, formerly a mathematician himself, tempted him by founding an Institute for Advanced Studies. Then it was back to Oxford, then a spell in Belgium, and finally Dublin once more in 1939, where, at long last, he settled for the next two decades, and had two more children by two different Irish women.

It was Schrödinger's complicated love life that precipitated his major discovery. Back in the halcyon days of his six-year sojourn in Zürich, he read a minor footnote in a paper by Einstein to the effect that de Broglie's ideas about electrons travelling along on pilot waves were 'probably more than just an analogy'. He was inspired to find out if that was true. In November 1925, he wrote to Einstein: 'A few days ago I read with great interest the ingenious thesis of Louis de Broglie, which I finally got hold of. It is extraordinarily exciting, but still has some very grave difficulties.' Just like de Broglie – indeed, almost simultaneously with him – Schrödinger thought he could explain those mysterious and arbitrary electron jumps from one orbit to another in Bohr's solar system model of the atom. But unlike de Broglie, he wasn't so keen on the idea that an electron might be 'the foam on a wave of radiation'. He thought the whole thing could be simplified into one component, the wave alone. At the comparatively late age of 38, his creative juices were brought to the boil by a passionate love affair during the Christmas holiday of 1925, conducted in the peaceful seclusion of a favourite hotel-sanatorium at Arosa in the Swiss Alps. His companion was a young woman from Vienna whose name has been lost to history. She was the anonymous muse behind one of the greatest ideas in the history of science.

Schrödinger's theory of 'wave mechanics' was comfortingly similar to classical equations developed earlier for describing wave phenomena in our more human-scaled world: water waves, sound waves, the vibrations of a violin string and so on. In fact he used that very analogy himself. The values describing Bohr's discrete electron energy levels 'occur in the same natural way as the integers specifying the number of nodes in a vibrating string. I believe [this] touches the deepest meaning of the quantum rules.' Think of a violin string resonating in such a way that the waves appear to stand still. Where an 'up' phase of a wave ends and a 'down' one begins, there is a point-like intersection on the string that appears to be motionless, especially if you become fixated by just that point in isolation. In fact it's just another part of the smoothly vibrating whole. The exclusion zones between energy levels (electron orbits) were not mysterious after all. They were just quiescent nodes in the wave, Schrödinger believed.

If the old bullet-like notion of particles had to be sacrificed, then still this 'wave function' had to represent something real. It couldn't be just a set of phrases in an equation. Schrödinger tried all kinds of mind's-eye visualisations. Perhaps the waves were made up of many smaller waves, and it was these condensed 'wave packets' that gave the superficial appearance of being bullet-like particles? Unfortunately, every time he attempted to convert this tidy mental model into mathematics, the whole scheme proved unstable. Then he tried thinking of waves as somehow a spread-out form of an electron. Again, this didn't quite work. Experiments with cloud chambers showed that free-flying electrons always left neat line-like droplet trails, just as might be expected if they were whizzing through the apparatus like hard little bullets. Those condensation trails never smeared out as if shunted by waves. Schrödinger had no way of explaining how his wave packets could remain so compact as they flew through space.

Yet his overall approach was preferred by many physicists as it was so much more beautiful than Heisenberg's. His differential equations for the wave functions made the atom seem almost tameable. And the waves could so *nearly* be visualised as something physical, something real. But nearly wasn't good enough. The problem with both de Broglie's scheme and Schrödinger's was that neither one had solved the mystery of electrons behaving like billiard balls in one experiment and looking like waves in another. Another problem was that a very annoyed Werner Heisenberg was ready with his counter-attack. And he had the great Niels Bohr on his side.

The Solvay wars

In 1911 a successful and socially progressive Belgian industrial chemist was keen to discuss his personal ideas about physics. Ernest Solvay had made a fortune by devising a method for making sodium carbonate ('washing soda') and now was in a position to pay for the world's great physicists to come and listen to his thoughts. In October he funded a five-day conference in Brussels. The delegate list was startling: Marie Curie was there, along with Max Planck, Ernest Rutherford, Henri Poincaré, Louis de Broglie and Albert Einstein, at that time 32 years of age and well respected among the science community at least, even if the world in general had barely heard of him yet.

First the delegates listened patiently to Solvay's presentation. They were prepared to be lenient with him, for he had been persuaded to fund what came to be known as the Solvay Conference as a regular forum in which physicists from around the world could exchange ideas. As Europe drifted towards war, the scientific community tried to resist the petty nationalisms and regional jealousies exhibited at that time by their various domestic political masters. Science should be international and free from politics, they believed. With an ebullient and ambitious Einstein leading the field, the first Conference in October 1911 was a fantastic success.

Of course the Great War ruined the atmosphere. When the conferences resumed in the 1920s, German scientists were barred, with the exception of Einstein, whose pacifist stance during the war and subsequent fraught relationship with the German authorities left plenty of room for manoeuvre. He nevertheless thought the ban was wrong, and argued vigorously that Britain and France's demands against Germany for war reparations were inhumane and unreasonable. He saw what so many others refused to see: the dangerous resentments that would trigger yet another war. He was absent from the 1921 Conference because it conflicted with a trip to America, and by 1924 he had made it clear that he would accept no more invitations until the anti-German prejudices had been removed. The organisers got the message, and German scientists were once more allowed to make their presence felt on the international stage. The way was cleared for one of the most incredible scientific gatherings in all of modern history: the Fifth Solvay Conference of October 1927. Here, the Young Turks of quantum mechanics would confront the last vestiges of the classical viewpoint and defeat it.

First there were some disagreements in their own camp that had to be resolved. The quantum mechanics faction was split along lines of its own: Bohr and Heisenberg, who believed that the atomic realm was intrinsically abstract and unknowable, and Erwin Schrödinger, who was equally convinced that his wave functions applied to something real. Schrödinger had written in *Annalen der Physik* that 'no genetic relation whatever with Heisenberg is known to me. I knew of his theory, of course, but felt discouraged, not to say repelled, by [his] methods of transcendental algebra, which appeared very difficult to me.' Heisenberg responded in an angry letter to Wolfgang Pauli. 'The more I reflect on the physical portion of Schrödinger's theory the more disgusting I find it. What he writes on the visualisability of his theory is, as Bohr might say, "probably not quite right". In other words, it's crap.' In July 1926 Schrödinger was invited to present his ideas at a lecture in Berlin, where he was met by an enthusiastic response, except from the young Heisenberg who stood up and complained. The packed audience booed him to sit down. He left the room despondent, and headed for the hills and the comfort of his youth group.

On 1 October that year Schrödinger arrived at Bohr's Institute in Copenhagen for an epic three-way row with him and Heisenberg. Actually Bohr himself collected Schrödinger from the railway station and the two men joined battle even before they'd left the platform. Schrödinger said: 'You surely must understand, Bohr, that the whole idea of quantum jumps necessarily leads to nonsense! [You say that] the electron jumps from one orbit to another one and thereby radiates. Does this transition occur gradually or suddenly? If the transition occurs suddenly, one must ask how the electron behaves in a jump? The whole idea is sheer fantasy!' Bohr replied that Schrödinger was too caught up in linguistic and traditional images of undulating waves. 'The pictorial concepts we use to describe the events of everyday life, and the experiments of the old physics, cannot represent the process of a quantum jump.' He never wavered from his belief that a conventional description of particles or waves operating in ordinary three-dimensional space and time was impossible. Schrödinger stuck to his guns. If quantum mechanics could not be 'fitted into space and time, then it fails in its whole aim, and one does not know what purpose it really serves'.

The arguments grew so intense that Schrödinger had to retire, exhausted, to his bedroom on the guest floor of the Institute. He realised he had made a big mistake by accepting an invitation to stay there, but itinerant theoretical physicists could not afford to turn down such offers. His host sat on the end of his bed, haranguing him further. 'Schrödinger, you must admit ... Surely you can see ...' At least Bohr's charming wife Margarethe had the decency to feed Schrödinger some soothing broth. The intensity of the discussions left both men drained. On being invited to visit Cambridge a few months later, Bohr wrote to Rutherford: 'I look forward to private discussions about our present theoretical troubles, which are of an alarming character indeed.' Rutherford wrote back in typically robust style. 'You cannot expect to solve the whole problem of modern physics in a few years. So be cheerful over the fact that there is still a great deal to do.'

The dispute had not been settled by the time of the 1927 Solvay Conference, where Heisenberg curtly destroyed all previous classical assumptions about the behaviour of matter. 'The more precisely the position [of a particle] is determined, the less precisely the momentum is known in this instant, and vice versa,' he told the delegates. This has been known ever since as 'Heisenberg's uncertainty principle'. Basically it means this. We can discover the energy of a particle at the cost of knowing exactly where it is in space; or we can isolate its position at the cost of learning about its energy. The more accurately one measurement is made, the hazier the other becomes. We can never obtain perfect information about a particle's position *and* its energy. It's as if a boxer can sense how hard his opponent's fist (a particle) slams into his body (a detector), but not where that fist is; or he can see exactly where the blow lands, but cannot tell if the impact is hard or soft. No ordinary language can convey the strangeness – the uncertainty – at the heart of the quantum world. When we choose which type of information we want to measure, the other information is lost. Reasonably accurate probabilities for a particle's behaviour can be calculated via quantum mechanics, and the larger the population of particles we're dealing with, then the more accurate the quantum predictions become. That's the best we can do.

It's easy to imagine that Heisenberg was saying something quite mundane: atoms and subatomic components are so small that no physical scientific instrument can obtain perfectly accurate information about them. The instruments themselves are made out of atoms, so perhaps they interfere with the atoms they are trying to investigate? But he was saying something much more profound. Even disregarding the party-crashing clumsiness of our instruments, the atomic realm is *intrinsically* uncertain. We can never know in advance when an alpha particle or a gamma ray is going to fly out of a radioactive atom. We can never mark, in advance, the exact spot on a scintillation screen where a particle will strike. Heisenberg told the Conference something even more mind-bending about the tracks and trajectories made by a particle as it sweeps through a scientist's instruments. 'I believe that the existence of the classical path [of a particle] can be formulated as follows: the path comes into existence only when we observe it.' This was one of the most shocking assertions in the history of human thought. We are used to a world in which events take place regardless of what we do. We might come along after an event and learn about what happened, or we might not; in which case the event has still happened anyway. Heisenberg claimed that the subatomic world doesn't work that way. Instead, our conscious observation of a particle *determines what has happened* to it. It was – and it remains today – the strangest idea in all of physics.

Albert Einstein wasn't sure about any of this. For him, the mathematics of quantum theory were convincing so far as they went, yet the very fact that they generated such a bizarre view of the world told him that the theory was incomplete. He was unwilling to surrender his grip on objective classical reality. A fortnight after the Conference he wrote: 'Quantum mechanics is certainly imposing. But an inner voice tells me that it is not the real thing. The theory says a lot, but it does not really bring us any closer to the secret of The Old One. I, at any rate, am convinced that He is not playing at dice.'

This is the root source of one of Einstein's most famous sayings, although in much the same way that Humphrey Bogart never speaks the line 'Play it again, Sam' in the movie *Casablanca*, Einstein never actually spoke the words, 'God does not play dice.' Even so, the phrase perfectly encapsulates his opposition to the idea of a universe ruled by chance and probability. Another much-quoted and equally unverifiable story has Niels Bohr saying: 'Einstein, stop telling God how to behave!' It sounds better than what he actually said. 'Don't you think caution is needed in ascribing attributes to Providence in ordinary language?' Anyway, in a letter to

Einstein he warned that physicists might simply have to learn to live with the contradictions and ambiguities. So long as all the mathematics worked, it should be 'possible for us to keep swimming between the realities'.

He and Einstein argued relentlessly throughout the 1927 Conference, although they never lost their tempers and seemed very much to enjoy their clashes. Every day Einstein said that there must be some as-yet undiscovered classical resolution to the problem, and every day Bohr would reply that there could be no such thing as 'independent reality'. He also reinforced Heisenberg's disturbing idea that physicists would have to come to terms with a wave description of the subatomic realm that works when you're dealing in wave experiments, and a particle description that works when you're dealing in particles, but there can be no reconciling the two. Bohr's notion of 'complementarity' (otherwise known as 'wave–particle duality') lies at the heart of quantum mechanics, wedded to its close cousin in peculiarity, Heisenberg's uncertainty principle.

Desperate to refute these nonsensical ideas, Einstein posed a series of *Gedankenexperiments*, 'thought experiments', at the Solvay Conference. One of the most useful ways of checking any scientific theory is to take it to extremes and work out what might happen. If the theory breaks down under stress then it probably isn't a very good theory. One day, for instance, Einstein argued with Bohr that quantum mechanics did not yield a proper explanation for what happens in Young's classic two-slit experiment if only single light photons at a time could be sent through the apparatus – a perfectly possible idea if an extremely dim light source were used. Each photon presumably would have to pass through both slits at once and interfere with *itself* for the cumulative effect to produce a wave-like interference pattern on the screen behind. Obviously there had to be an explanation for that, and Einstein hadn't heard one from the quantum camp. Bohr and his supporters believed that even a single particle had to be treated as essentially a smeared-out cloud of indeterminate possibilities, until the observations at the end of the experiment 'decided' exactly where it had struck the screen. Einstein rebelled at this idea, but for once Bohr did not mince his words. 'We are presented with a choice of *either* tracing the path of a particle *or* observing interference effects.'

This was the pattern throughout the Conference: a breakfast *Gedankenexperiment* from Einstein and a resolute refutation from Bohr. Paul Ehrenfest, a mutual friend, felt like a child anxiously watching the divorce of two equally beloved parents, even if the quality of the debate was thrilling. 'Like a game of chess, Einstein all the time gave new examples, while Bohr, from out of the philosophical smoke clouds, constantly searched for the tools to crush one example after another.' When the Conference came to an end, and Einstein was on the train from Brussels to Paris, he confessed to his travelling companion Louis de Broglie that he was feeling a little old and tired, and that maybe a younger man would be better suited to carry on the struggle for reason. But it was de Broglie who caved in while Einstein carried on the fight into his old age.

The 'Old One' defeated

Einstein had no specific religion and did not believe in the anthropomorphic kind of god that would have been familiar to most ordinary people of his generation, yet throughout his life he thought that the fundamental truths of existence could be discovered and expressed in a single neat set of natural laws. It was a kind of faith, certainly. These laws, deeper than Man's temporary and muddled intellectual discoveries, and perfect in their essential simplicity, could not conflict with each other. Therefore, if quantum mechanics came up with a wave description of the atom *and* a particle one, and these two ideas seemed to clash, then a third and deeper law of nature, a 'hidden variable' that had yet to be discovered, must surely exist to reconcile them.

The hardline quantum theorists disagreed. In a joint paper to the Conference, Bohr and Heisenberg insisted: 'Quantum mechanics is a complete theory, no longer susceptible of modification.'

The Bohr–Heisenberg camp declared victory, and their so-called 'Copenhagen Interpretation' of quantum mechanics was absorbed into the scientific mainstream, with complementarity and uncertainty as key elements. It was, and it remains, just an interpretation. As we'll find out later, there are still many vigorous arguments about its validity. It may have gained such prominence only because of the strong personalities of Bohr and Heisenberg, and the hesitancy of other scientists, few of whom were so well equipped to express ideas about the new quantum theory. The other difficulty is that modern historians cannot agree precisely what the exact terms of the Copenhagen Interpretation actually were. Bohr and Heisenberg had their own arguments, quite apart from their joint criticism of Schrödinger's waves. Our general understanding of the Copenhagen Interpretation is that whatever the fine details, it gave scientific legitimacy to the 'weirdness' of quantum theory.

Subsequent laboratory experiments kept the weird factor very much alive. Two months after that momentous 1927 Solvay Conference, a pair of American physicists, Clinton Davisson and Lester Germer, published in *Nature* magazine the results of a diffraction experiment using a beam of electrons, those supposedly bullet-like particles of negative electric charge. In principle, their set-up amounted to a variation on Young's classic experiment with light back in 1801, although the diffraction effect relied this time on a crystal barrier rather than a flat sheet with slits. When all the data was collated, they exhibited exactly the same kinds of interference phenomena predicted by quantum theory. Matter, it seemed, really did behave like a wave. The behaviour of particles could be described using exactly the same wave equations that described how light worked. Davisson wrote afterwards: 'It is rather as if one were to see a rabbit climbing a tree. Cats climb trees, so if only the rabbit were a cat, we would understand its behaviour perfectly.' The interpretations we bring to such strange results are still open to question, but the experimental data strongly suggests that the atomic realm might indeed be as strange as Bohr and Heisenberg claimed.

The results were all the more remarkable because Davisson and Germer had not at first deliberately set out to find the interference patterns. Their discoveries emerged almost as a sideline while they made investigations during a convoluted legal dispute between the General Electric and Western Electric companies about the design of an electronic vacuum tube. In Aberdeen, a physics professor called George Thomson made similar investigations, and eventually won the 1937 Nobel Prize with Davisson 'for their experimental discovery of the diffraction of electrons by crystals'. Curious, given that Thomson's father, the great J.J. Thomson, had discovered that electrons behave like particles ...

We need to jump forward in time for a moment. The practical difficulties of quite literally firing individual electrons through two slits and recording the results seemed, for several decades, beyond the realms of even the most subtle technicians. In the early 1960s Richard Feynman made a brilliant series of lectures to his students, one of which focused on the two-slit conundrum. 'We choose to examine a phenomenon which is absolutely impossible to explain in any classical way, and which has in it the heart of quantum mechanics. In reality, it contains the only mystery. But we should say right away that you should not try to set up this experiment. The trouble is that the apparatus would have to be made on an impossibly small scale. We are doing a thought experiment. We know the results that *would* be obtained because there are many other [types of] experiments that have been done.'

Even as he spoke, the practical problems were being solved. In 1961 Claus Jönsson of the University of Tübingen in Germany performed an actual double-slit test with a beam of electrons for the first time. It was heralded by scientists around the world as perhaps the single most beautiful experiment in the history of physics. The next and most daunting challenge, sending just one electron at a time through the experimental apparatus, was solved by Akira Tonomura and co-workers at the Hitachi company in 1989. As predicted by quantum theory, individual electrons interfered with themselves when passing through both slits and made interference patterns in the detectors.

What happens is so extraordinary it defies the imagination. If one slit is blocked off, each electron flies neatly through the open slit and delivers a single scintillation on the sensors behind, as if the electron were a compact little bullet. The accumulated scintillations form a neat shadow of the slit. On the other hand, if both slits are open, each electron seems to pass through both slits at once, interfering with itself and landing at some completely random and unpredictable point on the screen. The cumulative effect from multiple electron impacts is a fuzzy interference pattern on the screen, just as a wave would make. How can an electron 'know' in mid-flight whether it's going to encounter one or two slits? The latest experiments allow for (almost) no doubts. When electrons are fired just one at a time, even at intervals of many seconds, the same results are observed: wave-like interference patterns when firing through two slits, and no interference pattern when only one slit is open. What's more, if a detector is positioned behind one of the slits to check that electrons do indeed spread out and go through both slits at once, then we only ever see electrons choosing one slit, and the interference pattern is lost. 'Particle–wave duality' presents us with a disturbing conundrum. Our very act of observing a particle seems to affect, in advance, what it will do, and how it will appear to us.

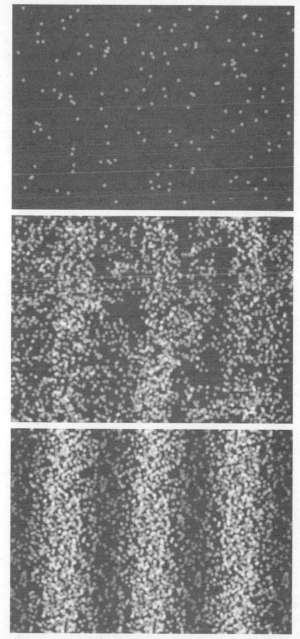

Figure 2. In a double-slit experiment performed with electrons, the pattern made by the first few strikes on the detector screen seems completely random, but as more and more individual 'hits' are recorded, a wave-like interference pattern emerges.

Spooky action

The two-slit problem was by no means the only shock. In 1935, by now securely installed at the Institute for Advanced Study at Princeton, New Jersey, Einstein continued his desperate struggle in support of a sane and ordered universe. He teamed up with two Institute colleagues, Boris Podolsky and Nathen Rosen, and wrote a paper which – he hoped – would reveal the gaping holes in quantum theory. It's been known ever since as the 'EPR paradox', after E-instein, P-odolsky and R-osen, naturally enough.

Suppose two particles are made to interact in an experiment, and then the scientists pause for some while before making an observation of a particular characteristic: say, the spin of one of those particles. Subatomic spin is a complex piece of behaviour, as we've seen, but not entirely dissimilar to the spin of billiard balls. For now, imagine the particles as just that – two billiard balls. They bash together, and one goes flying off, while the other hurtles into a measuring device, where the experimenters discover that it's spinning clockwise very rapidly. This means that when it was in contact with the other ball, some of that spin must have been transmitted back and forth between them, so that ball number two presumably has an anti-clockwise component in its spin. In the classical world, we wouldn't think twice about this.

Quantum billiard balls, of course, are different. According to quantum mechanics, each ball (particle) is somehow (and this is where it's impossible to visualise) spinning *both ways at once*. Observing a ball forces it to 'make up its mind', at which point a definite spin becomes apparent as a result of that observation. So, what about the ball that hasn't been measured – the one that's gone flying off into the distance? How does this second ball 'know' which spin to adopt now that the first ball's fate has been decided by the scientist's measurements? Einstein, Podolsky and Rosen argued that some eerie faster-than-light force, some 'spooky action at a distance', would have to be transmitted between the balls. Since Einstein had famously proved that no object or signal can travel faster than light, the idea of instantaneous communication between the particles was out of the question. Something was wrong with the uncertainty principle. In May 1935 the three collaborators published their paper, 'Can Quantum-Mechanical Description of Physical Reality be Considered Complete?' It was a rhetorical question. In their concluding remarks, they held out for a return to normality. 'While we have shown that [quantum mechanics] does not provide a complete description of physical reality, we left open the question of whether or not such a description exists. We believe, however, that such a theory is possible.'

Again we have to jump forward in time for a moment. In 1982 a team at the University of Paris, led by the physicist Alain Aspect, ran an Einstein–Podolsky–Rosen experiment for real. They measured the polarisation of light photons: the spatial components of their quantum numbers (in other words, the wave-like 'orientation' of photons!). And apparently, it worked. No matter how great the distance between them might be, 'a pair of entangled [particles] should be considered as a global, inseparable quantum system', Aspect concluded. In other words, the 'spooky action at a distance' might just be happening. As always with quantum ideas, some interpretations differ. But the essential *puzzle* simply deepens.

The dead-and-alive cat

Back to 1935, when EPR still seemed like an argument *against* quantum mechanical peculiarities. In the nine years that had passed since Schrödinger had first tried to defend himself against Heisenberg and Bohr's maddening abstractions at Bohr's Institute in Copenhagen, he had lost none of his faith in an underlying physical reality. He was electrified by the Einstein–Podolsky–Rosen paper, and eagerly contributed his own thoughts in support. It was all very well highlighting the absurdities in subatomic descriptions of matter, but what would happen if quantum mechanics extended to things one could actually see and touch? Everything is made from atoms, after all. At what point do the peculiarities of quantum mechanics impinge on what we think of as the real and everyday world? 'Serious misgivings arise if one notices that the uncertainty affects macroscopically tangible and visible things,' he wrote. His answer was the famous problem known as Schrödinger's Cat:

> One can even set up quite ridiculous cases. A cat is penned up in a steel chamber, along with the following device (which must be secured against direct interference by the cat): in a Geiger counter there is a tiny bit of radioactive substance, so small that perhaps in the course of an hour one of the atoms decays, but also, with equal probability, perhaps none decay; if it happens, the counter tube discharges and through a relay releases a hammer which shatters a small flask of hydrocyanic acid. If one has left this entire system to itself for an hour, one would say the cat still lives if meanwhile no atom has decayed. The ... system would express this by having in it the living and dead cat mixed or smeared out in equal parts.

To put it even more simply, one couldn't have subatomic particles that were neither one thing nor the other, unless one also had *things* that were neither one thing nor the other. If taken to its logical conclusion, quantum mechanics suggested a world so much at odds with human understanding that it simply could not make sense: a world in which cats could be both dead and alive at the same time. Heisenberg had told the 1927 Solvay Conference that 'the path [of a particle] comes into existence only when we observe it.' Who, or what, did he mean by 'we'? Could the cat inside the box be its own observer? If so, then what other creatures might go into the box instead? A mouse? An ant? A single-celled amoeba?

Nor does opening the box after the hour is up help eliminate the weirdness. Yes, the cruel experimenter will discover a cat that is very much alive, or one that is quite certainly dead, but ultimately this is just an elaborate experiment to observe whether or not that tiny little speck of radioactive material, with its 50–50 probability of discharging its particle, has actually done so. The aliveness or deadness of the cat is, so to speak, just a read-out of the experiment's results. And since the path of the radioactive particle is determined only by the observer at the end of the experiment, so the cat's aliveness or deadness is also determined by the experimenter when he or she opens the box ... There is no independent reality to rely on. Schrödinger thought this was nonsense, and that the world must surely get on with the business of being real 'out there' (or in a box) regardless of human observation. If all of us disappeared tomorrow, the planets would still tumble in their orbits, the stars would still shine, living cats would continue to live, and ones fated to die would die anyway. The radioactive pellet must do whatever it will do, regardless of whether or not we humans come along to observe it afterwards.

We still don't know what any of this means. The distinguished physicist John Wheeler has neatly described the ultimate mystery of quantum physics. 'May the universe in some sense be "brought into being" by the participation of those who participate? "Participation" is the new concept given by quantum mechanics. It strikes down the "observer" of classical theory, the man who stands safely behind the thick glass wall and watches what goes on without taking part.'

That could easily be a description of our next character, Paul Dirac, the brilliant and troubled man who proved that Heisenberg and Schrödinger's mathematics were essentially the same – even if they argued about what it all meant.

Why do you dance?

There are so many novels and movies in which exaggeratedly monstrous parents inflict cruelty on children, and the entire cast of characters skulks around the shadowy staircases and dripping cellars of loveless and doom-laden homes. Most of the time we enjoy these tales because they remind us that our own families, although far from perfect, at least aren't *that* bad. The early life of Paul Dirac shows that even the darkest gothic nightmares are inspired by real-life possibilities. Charles Dirac was an autocratic French teacher at a secondary school allied to the Merchant Venturers Technical College in Bristol. He had few friends, and no interest in broader society. When he came home from work in the evening he closed the front door of the modest Dirac house and shut out the world.

Born in 1902, Paul was one of three children forced to live under Charles Dirac's grim regime. His father insisted, for instance, that everyone spoke only French when sitting down for supper. Paul tried his best, and was allowed to sit in the dining room while the rest of the family, Paul's mother Florence, older brother Reginald and younger sister Beatrice, ate in the kitchen. Little wonder that Paul grew up pathologically shy and tongue-tied almost to the point of complete silence. The strange thing is that Charles himself had run away from an unhappy childhood in Switzerland, yet this so-called teacher had learned no lessons from his experience. He was the embodiment of Philip Larkin's warning that 'Man hands on misery to man / It deepens like a coastal shelf.' As Charles' two sons came of age, he forced them to study engineering. Reginald committed suicide at 24, but Paul escaped to Cambridge where he found refuge in theoretical physics. The mathematical abstractions of the new quantum theory entranced him even as he shied away from the terrors of the human world. He once wrote: 'One perhaps could say ... that God is a mathematician of very high order, and he used very advanced mathematics in constructing the universe.'

Unlike so many other physicists, who seemed to have an irrepressible need to swap ideas, argue their pet theories and shoot down the ones they disagreed with, Dirac shunned human company and almost always worked in complete isolation. Of the 250-odd scientific papers and other publications he produced in his long career, only a handful are co-authored. He relaxed by taking long walks on Sundays, and, later, by extensive foreign travelling, again usually on his own. For many years he seemed to have no idea how to make relationships. The sexual world was not only alien to him, it was unfathomably puzzling. On one especially poignant occasion in 1929, Heisenberg and Dirac were invited to lecture in Japan, and travelled there aboard an ocean liner. There were regular social events and dances. Heisenberg would always enter into the party spirit, while Dirac would sit quietly in a corner – if he even came at all to the dances – and watch. One night he asked Heisenberg: 'Why do you dance?' Heisenberg answered, naturally enough: 'Well, when there are nice girls around, then I feel like dancing with them.' Dirac fell silent. It was only several days later, after giving the matter considerable thought, that he called Heisenberg on his cabin phone and asked his next question. 'How do you know beforehand that the girls are nice?'

The mysteries of women aside, Dirac's great insight was that Schrödinger and Heisenberg were essentially describing the *same* behaviour for electrons, even if one man relied on abstract algebraic systems and the other visualised some kind of a wave. He created a mathematical bridge, simplifying the terms, reviving some techniques of classical mechanics dating from the 1830s. He also neatly incorporated aspects of Einstein's relativistic energy–mass equations, taking into account that particles must gain mass as they accelerate. Even at half the speed of light, the mass calculations for any particle veer significantly from classical expectations. At 80 per cent lightspeed, a particle's mass increases by more than 60 per cent, and so on. Within a few years of Dirac's accommodation of relativity into atomic science, machines would emerge capable of boosting particles to substantial fractions of the speed of light ...

Dirac may have been shy in company but he did not lack intellectual confidence. Once, in the earliest phases of his work, he sent a 30-page draft of his ideas to Heisenberg, who wrote back saying: 'I have read your extraordinarily beautiful paper on quantum mechanics with the greatest interest, and there can be no doubt that all your results are correct.' Max Born was similarly impressed. 'The name Dirac was completely unknown to me ... the author appeared to be a youngster, yet everything was perfect in its way, and admirable.'

Matter in the mirror

Dirac's equation did rely on some unusual assumptions. There were two possible solutions, one describing the behaviour of electrons, and the other yielding a particle with opposite electrical charge, an anti-electron or 'positron'. Stranger still, these mirror image particles, known as 'anti-matter', would possess negative energy and negative mass. These were difficult ideas for anyone to grasp.

In 1932 an American physicist, Carl Anderson, was monitoring energetic particles that seemed to rain down on earth from space, the so-called cosmic rays. Some of the tracks in his cloud chamber equipment seemed similar to electrons except that they were deflected the wrong way by the chamber's surrounding magnets. They even flew the wrong way through the system, as if climbing back towards the sky. This was the first known detection of anti-matter, and it soon became clear that the positron is not some exotic rarity. Dirac showed that no component or sub-component of matter is without its shadowy counterpart in the anti-world.

According to our most modern theories, the Big Bang that created our universe should have created just as much anti-matter as matter. So how come our universe still has so much matter left in it? If particles outnumbered anti-particles in the Big Bang by as little as one part in 100 million, then the existence of our universe could be due to the tiny fraction of particles that had no anti-particles to wipe them out. Other theories suggest that even if identical amounts of anti-matter and matter were created in the Big Bang, their respective physics may have been slightly different. This difference might have favoured the survival of matter, for some mysterious reason. On Earth, anti-matter is routinely harnessed. Low-energy positrons, the products of the natural decay of certain radioactive isotopes, are used in medical imaging. Positron Emission Tomography (PET) follows the progress of a weakly radioactive positron-emitting tracer chemical injected into the bloodstream. When the tracer reaches active areas of the brain, the positrons knock into nearby electrons, generating gamma rays that can be detected by the scanner. PET allows us to monitor areas of the brain that become activated during particular kinds of thought process. Low-energy anti-matter helps us learn more about the mind that was capable of discovering anti-matter in the first place. Higher-energy anti-matter particles are produced at only a few of the world's largest particle accelerators. The current worldwide production of anti-matter amounts to a few millionths of a gram per year.

Dirac's final challenge was to try to describe not just the behaviour of atoms and electrons on their own, but what happens when electromagnetic energy and matter interact. Time and again we have seen how important is the connection between photons of light energy and the electrons in an atom. An absorbed photon kicks an electron into a higher energy state (or 'orbit'). An electron dropping suddenly from a higher state to a lower one will emit a photon, and so on. All the light we see around us is generated by photons kicked out of atoms by electrons dropping from higher to lower energy levels. The forms of light we *don't* see

– infra-red, ultraviolet, x-rays and gamma rays – are created in the same way. Dirac tried, and failed, to understand more about this process, known as Quantum Electrodynamics or QED. In the meantime, he was forced into greater social openness by his increasing fame among the physics community, not to mention the joint award of the 1933 Nobel Prize for Physics to him and Heisenberg 'for the discovery of new productive forms of atomic theory'. He even managed to get married in 1937 to Margit Wigner (the sister of another well-known physicist, Eugene Wigner). Margit had two children from a previous marriage, and the couple had two more children of their own. Dirac achieved some measure of happiness, yet he could never entirely free himself from his tragic past. Margit wrote in a personal reminiscence: 'Paul, although not a domineering father, keeps himself aloof from his children. That history repeats itself is only too true in the Dirac family.'

PART FOUR

Playing with Marbles

Early experimenters believed they could crack open the secrets of the atom by bombarding it with other particles. Their aim was merely to unlock further secrets of nature. Inadvertently they stumbled on the most dangerous discovery that science has ever made.

George Gamow was born in the Ukrainian port city of Odessa in 1904. His family was pretty colourful, at least on his mother's side, although George had to rely largely on his father's accounts to discover this, as his mother died while he was just a toddler. There was an adventurous battleship commander in the Gamow clan, and a somewhat less successful astronomer who was hanged for plotting to assassinate the Russian Prime Minister. There was an Orthodox archbishop in the mix too. He would hardly have approved of young George's teenage experiments with microscopes and Holy Communion wafers. He was looking for evidence of the transubstantiation of bread into the body of Christ. 'I think this was the experiment which made me a scientist.'

The ill-equipped local university at Odessa was no fit place in the 1920s for a bright young man. Gamow's father sold all the worthwhile possessions he had, including the family silver, and sent George to Petrograd where he could pursue his studies in physics. He paid his way by teaching basic science to Soviet Red Army cadets, meanwhile pretty much coasting through his undergraduate work and skipping most of his lectures. He preferred instead to hang out with a coterie of like-minded bright slackers nicknamed the Jazz Band. Even as the city around them was renamed Leningrad, they seemed unconcerned at the dangers of flaunting their Bohemian style, watching silent movies, playing tennis and childish parlour games, and drinking in dubious dens. Lenin and his fast-rising lieutenant Stalin were keen to repress the slightest signs of bourgeois rebellion, yet also determined to hasten the modernisation of the Russian empire's old-fashioned and largely rural economy. They wanted factories, steel mills, machines, hard-edged technology.

The Soviets had dismantled the Tsarist trappings of luxury and privilege, but some of the institutions left over from the days when Russian academic life was inspired by the Napoleonic French example still survived. The Academy of Sciences, for instance, maintained many of its old privileges. Science was one of the main tools with which the Soviet leadership hoped to forge a new and modern Socialist paradise. And so the strange circumstance arose in which Russian scientists were allowed the freedom to travel abroad and soak up new ideas from foreigners, even in the very heart of the capitalist West.

In August 1928 George Gamow arrived penniless and homeless at the Niels Bohr Institute in Copenhagen. He had spent a productive year in Göttingen absorbing the latest theories from the champions of quantum mechanics, and now he wanted to explain to Bohr his startling new explanation for radioactive decay. It had to do with those sudden quantum 'jumps' that Bohr had described. So far, Rutherford and other investigators had been fooled, as it were, by the apparent violence with which alpha particles were thrown out of a radioactive nucleus. Rutherford's muddled explanation for this had confused everyone, not least himself. Gamow realised that the quantum theorists had been busy looking at how orbiting electrons behave, but because they weren't really experimentalists, they hadn't

1. A youthful Marie Curie in the 1890s, at the time of her greatest happiness with Pierre Curie. (AIP Emilio Segrè Visual Archives, W.F. Meggers Collection)

2. Pierre and Marie Curie in a stiffly formal pose that reveals little of the great love they felt for each other. (AIP Emilio Segrè Visual Archives)

3. Max Planck, the conservative theorist who triggered an unwelcome revolution in science when he discovered the 'quantum of action' in 1900. (Photograph by Maison Albert Schweitzer, courtesy AIP Emilio Segrè Visual Archives)

4. Albert Einstein in an early portrait that reveals his mischievous nature. (© Bettmann/CORBIS)

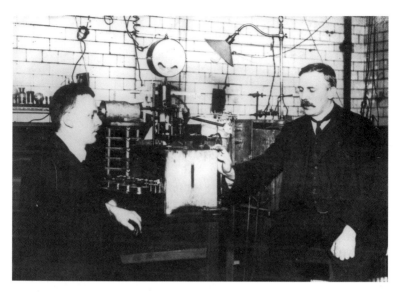

5. Hans Geiger (at left) and Ernest Rutherford at the University of Manchester in 1908. (AIP Emilio Segrè Visual Archives, Physics Today Collection)

6. The two Grand Lords of British physics, J.J. Thomson and Ernest Rutherford. (Photograph by D. Schoenberg, courtesy AIP Emilio Segrè Visual Archives, Bainbridge Collection)

7. Niels Bohr, the 'Great Dane', became a trusted father-figure to the quantum generation. (AIP Emilio Segrè Visual Archives, W.F. Meggers Collection)

8. Wolfgang Pauli's friends and colleagues appreciated him as a
   brilliant and acerbic sparring partner who took no prisoners in any
   argument. (© CERN, Geneva)

9. A youthful Werner Heisenberg, not yet clouded by the atom's darker potentials. (AIP Emilio Segrè Visual Archives, Segrè Collection)

10. Erwin Schrödinger, the sensualist inspired by a 'late-flowering erotic outburst' into a new appreciation of the atom. (AIP Emilio Segrè Visual Archives)

11. The Fifth Solvay Conference of 1927, arguably the greatest gathering of physicists the world has ever seen. Among many great names, Erwin Schrödinger is in the back row, with Wolfgang Pauli and Werner Heisenberg. In the middle row are Paul Dirac, Louis de Broglie, Max Born and Niels Bohr. The front row boasts Max Planck, Marie Curie, Albert Einstein, Marie's illicit lover Paul Langevin, and cloud chamber inventor Charles Wilson.

12. Paul Dirac, the shy Englishman who reconciled Schrödinger and
    Heisenberg's rival theories. (Photograph by A. Bortzells Tryckeri,
    courtesy AIP Emilio Segrè Visual Archives)

13. John Cockcroft (left) and George Gamow compare notes at the
    Cavendish Laboratory in 1930. (AIP Emilio Segrè Visual Archives,
    Bainbridge Collection)

14. Ernest Walton, Ernest Rutherford and John Cockcroft celebrate 'splitting the atom' with the world's first linear accelerator in 1932. (AIP Emilio Segrè Visual Archives)

15. James Chadwick, whose subtle experiments revealed the neutron's presence inside the nucleus. (Photograph by Bortzells Esselte, Nobel Foundation, courtesy AIP Emilio Segrè Visual Archives, Weber and Fermi Film Collections)

16. Otto Hahn and Lise Meitner's scientific partnership was politically and emotionally complex. (AIP Emilio Segrè Visual Archives, Brittle Books Collection)

17. Meitner's nephew Otto Frisch coined the term 'nuclear fission'. (AIP Emilio Segrè Visual Archives, Segrè Collection)

18. Enrico Fermi, the Italian-born physicist who built the world's first nuclear reactor underneath an abandoned Chicago sports arena. (AIP Emilio Segrè Visual Archives)

19. Robert Oppenheimer, the troubled 'father' of America's atomic
bomb. (Los Alamos Scientific Laboratory, courtesy AIP Emilio
Segrè Visual Archives)

20. Edward Teller (seen here, at left, with Enrico Fermi) was the real-life inspiration for *Dr Strangelove*. He was an unapologetic champion of nuclear weaponry. (AIP Emilio Segrè Visual Archives)

21. Murray Gell-Mann (left) and Richard Feynman, great friends and perhaps even greater rivals. (AIP Emilio Segrè Visual Archives, Marshak Collection)

22. Freeman Dyson played the 'Dirac' role by finding common ground between Feynman's neat diagrams and Schwinger's tangled equations. (AIP Emilio Segrè Visual Archives)

23. Fred Hoyle worked out how the atomic elements are created in stars, but refused to believe in Big Bang theory. (AIP Emilio Segrè Visual Archives)

played around much with the nucleus. What, he argued, if the nucleus could also be described in quantum terms? If an electron could jump from one orbit to another without occupying any intermediate half-and-half orbits, then could an alpha particle jump (or 'tunnel') just as suddenly from being inside the nucleus to being outside of it? Then the charges on the newly isolated particle would be repelled by like charges in the remaining nucleus, and the orphan fragment would hurl across space *as though* it had been violently ejected. But Gamow argued that the actual business of pushing it through the door and out of the house, so to speak, never really happened. Radioactive decay was just another sudden quantum jump from one state to another.

Bohr listened, then straight away signed Gamow to the Institute for a year's work. Gamow thought he had landed in physicist's heaven. Barely four weeks after receiving such a warm welcome, he had the shock of his life. In the latest copy of the weekly London science journal *Nature*, he read with a sickening sensation the following words: 'Much has been written of the explosive violence with which the alpha particle is hurled from its place in the nucleus ... one would rather say that the alpha particle slips away almost unnoticed.' Quite independently of Gamow, two Princeton University researchers, Ronald Gurney and Edward Condon, had also worked out that Schrödinger's wave equations might be used to describe radioactive decay, literally by smearing the distinction between 'inside' and 'outside' the nucleus into a wave of probability rather than a hard boundary. Gamow had been beaten to the post. Bohr acted swiftly to ensure that his new young recruit would still have a chance to publish something worthwhile before all his thunder was stolen. But if he'd nearly lost that race, his most dramatic contribution now would be to turn his earlier ideas on their head. How could you get a particle *into* the nucleus?

The quantum mechanics in Europe had focused their attention almost entirely on electrons and photons.

Meanwhile, ever since that incredible day in the spring of 1911 when alpha particles had rebounded from gold foil like 15-inch shells bouncing off tissue paper, Ernest Rutherford had tried to learn more about the nucleus, his 'gnat in the Albert Hall'. Progress had eluded him. The great powerhouse New Zealander appeared to be losing his boom. Part of the problem, he believed, was that the instruments in his laboratory were not sufficiently powerful. The naturally occurring radioactive sources for alpha and beta particles were, by and large, too weak and random in their emissions. They could not easily be calibrated as a precise instrument for bombarding targets. The whole business of counting scintillations in the dark was subject to human error, human tiredness. It was all too easy to miss important scintillations, or to imagine sudden tiny sparks in the dark that had no reality except in the overworked mind's eye.

And anyway, the nuclei simply refused to yield up their secrets. They just wouldn't break. In 1919 Rutherford did find something incredible when he bombarded nitrogen gas with alpha particles, even if it was not quite what he expected. He 'transmuted' nitrogen into different elements by splitting the nitrogen's nucleus, knocking out protons (the carriers of positive charge in the nucleus) and forcing the alpha particles to combine with the nitrogen fragments left behind, making oxygen and hydrogen. The results were too sporadic to be of much further use. He could find no consistently reliable way of splitting nuclei to reveal more of their secrets. The papers from his lab began to adopt a somewhat forlorn tone. 'Unfortunately the amount of disintegration [is] very small … Unfortunately, on account of the small number of particles …' Two years later, the great J.J. Thomson resigned as head of the Cavendish Laboratory to become the Master of Trinity College. Rutherford agreed to replace him, initially reluctant as he thought that Thomson must surely want to keep control from the sidelines. J.J. assured Rutherford that he would have a free hand.

By 1927 Rutherford was actually starting to lose his nerve

about cracking open the nucleus. Some of his lectures became, if not incoherent, then certainly unconvincing. His attempt, that year, to explain how alpha and beta particles were emitted by radioactive substances such as radium and uranium were so muddled that one of his presentations actually made his loyal audience squirm in discomfort. In 1929 Bohr despatched Gamow to the Cavendish with his intriguing new ideas about how to tamper with the nucleus.

Big science

Seventy years ago the Cavendish was one of the most renowned physics laboratories on earth. 'Oh, we're just playing with marbles', Rutherford would joke, all the while in fact being immensely proud of what he saw as the real work of physics: actually exploring the behaviour of physical things. 'All science is either physics or stamp collecting,' he famously remarked. Heisenberg, Schrödinger and Einstein may have produced brilliant theories, and Rutherford certainly absorbed them, but at the same time he wished he could dispense with some of their wilder abstractions. Whenever a new quantum mechanics scheme emerged from the Continent, he'd mutter: 'Watch out, the thinkers are on their hind legs again.' And if any of his young assistants began to speculate about the big questions, such as the ultimate origins of existence, he'd complain: 'I don't want to hear any talk about the universe in *my* department!' He was in love with his practical devices, the vacuum tubes, cloud chambers and scintillation counters and all the other bench-top machinery pioneered at the Cavendish. Yet none of these tools could solve the biggest problem. The nucleus could be chipped, but still it would not break.

Rutherford decided that if he was going to try to smash nuclei apart to study their sub-components, he needed to hit them so damned hard that his 'trusty right hand', the alpha particle, wouldn't simply rebound off them. Given the positive charge of alpha particles (helium nuclei) they could be steered by electric and magnetic machinery. Could they perhaps be boosted and accelerated? Yes, but it would take a great deal of electrical power: far more than any battery or mains electrical supply commonly available.

In November 1927 Rutherford attended a meeting of the Royal Society at Burlington House in London, where he announced his idea for an incredible new scientific instrument. Its task was to generate a massive electromagnetic field that could repel freshly-manufactured charged particles and catapult them towards their targets at significant fractions of the speed of light. Thinking initially about electrons, he had in mind a device somewhat like J.J. Thomson's cathode ray tube, but on a much grander scale. As usual, a difference in the electrical potential between a cathode at one end and an anode at the other would cause electrons to come streaming across the gap. They could then be harnessed and steered by electromagnets and used as high-velocity bullets to smash targets. Variations on the device would accelerate alpha particles or protons (ionised hydrogen atoms stripped of their electrons) injected into the tube. But the magnitude of the electrical energies Rutherford had in mind was staggering. He wanted a power source delivering up to 8 million volts' worth of difference between the plates. This was in an age when half the houses in Britain were still lit by gas lamps, so there was no question of simply plugging such a machine into the wall socket. For the first time in its 60-year history, the Cavendish wanted to design a scientific instrument that was entirely beyond the scope of its hard-working laboratory technicians alone to build out of wood and glass and copper.

As soon as Rutherford had announced his grand ambitions, a junior colleague, the young Irish-born physicist Ernest Walton, tried to produce a flow of super-fast electrons. He thought he could dodge the need for millions of volts by exploiting just the modest electrical energies already available from the Cavendish's generators and battery accumulators. The most these could come up with was about 300,000 volts, but Walton's clever idea was to accelerate his electrons not just once but thousands of times in succession by sending them around a ring-shaped track, all the while suspended in an electromagnetic field. At the same time they would be driven forward by the rapid-cycling polarity of the field. The electrons would pick up an extra kick of speed every time they made a circuit. Walton was on to an important idea, but he couldn't work out how to keep fine control over the magnetic containment coils surrounding his device. The electrons smashed into the walls almost as soon as they entered the ring.

In 1928 Walton and his colleague John Cockcroft gained Rutherford's approval to build a 'linear accelerator' so that they could try to smash nuclei with proton bullets. The lab turned to private industry for help in developing this more powerful equipment. The nascent electrical industry was far from becoming the 'national grid' we take for granted today. Nevertheless, a serious effort was being made to wire up a greater proportion of domestic homes, and of course to power industries more efficiently. A 1926 Act of Parliament created a new entity called the Central Electricity Board. The Metropolitan-Vickers company in Manchester was one of the main suppliers of huge generators, capacitors, transformers, electromagnets and the other heavyweight paraphernalia of the electricity business. 'Met-Vick' even had its own high-voltage research laboratory. Rutherford and his team at the Cavendish drew increasingly on the expertise of these industrialists. The Royal Society made an impressive grant of some £15,000 to expand the Cavendish's somewhat gothic Victorian pile of a building so that it could accommodate the new equipment. The Department of Scientific and Industrial Research, a government arm established towards the end of the 1914–18 war, also contributed a number of grants. The era of 'big science' had arrived, and with it came an increasing burden of administration for Rutherford, which he did not welcome.

By 1930, as the accelerator machinery began to take

shape, George Gamow had joined the Cavendish for a year's visit, strongly encouraged by Bohr. The blond-haired Ukrainian with thick-lensed glasses and a high-pitched and chaotic multilingual way of talking impressed Rutherford for all the wrong reasons. His ramshackle British-made BSA motorcycle threatened to kill any number of Cavendish people, not by running them over but by tipping them off, as they lined up enthusiastically to try to ride the thing. Bohr also paid a brief visit and, spurred on by Gamow, took a turn on the BSA during a Sunday tea at the Rutherfords' house, bringing traffic to a halt as he skidded the bike. Mary Rutherford was unimpressed, and her husband likewise was less than thrilled.

Gamow's theories did not at first make quite so strong an impression as his mischievous personality. The Cavendish experimenters took on board his idea that a quantum jump might just as easily insert a particle into a nucleus as eject it. They knew, in principle, that it might take nothing more than, as it were, nudging a particle alongside a nucleus and waiting for it to make that jump from outside to in, at which point something interesting was bound to happen. They knew that if Gamow was right, this should be achievable at quite low voltages. And yet they clung to their dream of hitting the nucleus with colossal force. Rutherford had told the 1927 meeting of the Royal Society that the Cavendish hoped one day to explore aspects of nature that 'far outstrip our puny experiments in the laboratory. What we require is an apparatus to give us ten million volts which can safely be accommodated in a laboratory.' For now, there was the prospect of reaching only 300,000 volts with Walton and Cockcroft's proton accelerator. It would be enough, Rutherford believed, to obtain a few encouraging results and get the new technology moving in the general direction of the mega-voltages he wanted.

It took more than two years for Walton and Cockcroft to build their accelerator. In April 1932 it was ready. There was a roomful of strange equipment and a large table covered in dials and switches, so that the whole scene was strikingly reminiscent of a Boris Karloff science fiction fantasy. But the heart of the machine would have been recognisable to any cathode ray tube experimenter from the late Victorian era. It was essentially an eight-foot-long glass tube, mounted vertically, and with a large wooden box at floor level into which the experimenters could crawl to monitor their instruments.

Rutherford was impatient by now. He had invested a great deal of his laboratory's prestige on the accelerator. He was aware, too, that experimenters in the United States, led by Ernest Orlando Lawrence at Berkeley in California, were on the verge of catching up. One day in early April 1932 he stormed into the lab and flung his wet overcoat onto a convenient nearby protrusion. It was a live electrical terminal, and he promptly gave himself a nasty shock. His temper now at boiling point, he shouted at everyone to 'get those protons moving!' His ire was a little unfair. Although it was fully constructed, the accelerator had to be calibrated before any of its results could make sense, and its complicated scintillation counters and detectors needed more work. Rutherford wanted to see the machine in action immediately, so Walton and Cockcroft installed an old-fashioned zinc sulphide screen in a single afternoon. This technique was so antiquated by now that Walton had never even seen one before.

On the morning of 14 April, Walton made the first operational run. At one end he placed a lithium target, which he then bombarded with protons, pushed along by a cautious starting voltage somewhat less than the machine's full capacity. Crawling across the floor of the darkened laboratory and squeezing himself into the little observation box at the base of the accelerator stack, Walton put his eye to the microscope and was amazed to see the little zinc sulphide screen aglow with dozens of little sparks. Then hundreds. Then so many there was no possibility of even guessing how many flashes there were. Walton was so surprised, he crawled out of the box and groped his way to the control table on the other side of the darkened lab and switched the entire machine off. Then he went back into the box, just to check that the screen had gone dark again. It had. He repeated his crawl, switched the system back on, checked the screen. Once again it was aglow with uncountable scintillations. This was an unimaginably stronger result than Walton or anyone else had expected from the accelerator, especially at such minor voltages.

Rutherford was summoned, and squeezed his by now somewhat portly bulk into the observation box. His deep voice thundered across the room: 'Switch off the proton current! Increase the voltage! Switch on the current! Off again! On!' And so on, until he, too, was satisfied that the machine had worked beyond his wildest hopes. 'Those scintillations look mighty like alpha particles to me,' he said. 'I should know an alpha particle scintillation when I see one. I was in at the birth of the alpha particle!' Rutherford's habitual troop-rallying cry of 'Onward Christian Soldiers!' was untempered by the sign which his colleagues had suspended from the ceiling of the lab: *Talk Softly Please*.

The atomic arithmetic was easy to work out from the traditional periodic table of elements. The protons were essentially hydrogen atoms, minus the negligible mass of their electrons. The protons, therefore, had an 'atomic number' of 1. Lithium's atomic number was 3. An alpha particle (helium nucleus) with its pair of protons was 2. Protons were adding themselves to the lithium nuclei, then *splitting* them into two alpha particles. And all with the unexpectedly low power input of barely 200,000 volts. Unexpected, that is, by everyone except George Gamow, now back in Russia and unaware, for the moment, of this new development. It wasn't brute power so much as the sheer *number* of protons, a hundred million million per second, that had done the trick. The accelerator had thrown the quantum dice again and again so that thousands of protons had made their magical jumps into lithium nuclei, even if so many millions more had simply bounced off their targets.

The nucleus had been split. Rutherford swore his people to secrecy. It was vital, now, to publish something in advance of the Americans. A hasty five-paragraph letter to the weekly London magazine *Nature* was prepared for publication: 'Disintegration of Lithium by Swift Protons.' Once Rutherford was satisfied that the letter was scheduled to hit the presses ahead of anything the Americans might publish, he made an announcement at his favourite venue, the Royal Society, on 28 April. Immodest though he often was, he made sure on this occasion that Walton and Cockcroft received all due credit.

The first mass-market newspaper to run the story for public consumption was a trivial ragsheet called *Reynold's Illustrated News*. Its headline for Sunday 1 May 1932 read: SCIENCE'S GREATEST DISCOVERY. The somewhat loosely informed text said that 'the dream of scientists has been realised. The atom has been split.' The more serious newspapers took up the story with equal enthusiasm. The public knew very little about atoms, but they did understand that these were the fundamental building blocks of all existence. If human guile could break them, this was god-like work indeed.

Rutherford thought that all this talk of 'splitting atoms' was trivial. He and his fellow researchers had been splitting them for years. Every time they made an ion by shredding electrons, they were essentially splitting an atom. No, it was the splitting of the *nucleus* that counted. And there was something more: something in the voltages. Recall that Walton and Cockcroft's machine had hurled protons at a fairly modest voltage. When the Cavendish team at last managed to install some properly calibrated detectors at the bottom of the stack, they discovered something momentous. They were shocked by how far the liberated alpha particles penetrated into a Wilson cloud chamber once they had flown out of the shattered lithium nuclei. Whenever a single proton successfully hit a lithium nucleus at 200,000 volts, two alpha particles rushed out as if propelled by 8 million volts each: a combined total, per lithium atom, of 16 million volts.

We have to switch once again to a human-scaled picture. Imagine dropping a bowling ball (one of the protons in the accelerator beam) gently into a sack containing three more bowling balls (the trio of protons in the lithium nucleus). You would be surprised if the sack suddenly burst apart with incredible violence and two pairs of balls came hurtling out at supersonic speed and smashed into the walls of your bowling alley, embedding themselves deeply into the brick and plaster. This is what the liberated alpha particles had done.

The Cavendish team worked out that the disintegration of the lithium nucleus into two alpha particles accounted for almost all the original mass. But not quite. Two per cent of the mass had vanished. This, they realised, had been converted into the energy required to throw those alpha particles out of the nucleus with such staggering force. It was as if an unimaginably powerful pent-up spring of energy had been released. It was the first practical demonstration that Albert Einstein's famous equation, $E = mc^2$, was correct. Matter was indeed an incredibly compacted form of energy, and the compression factor was the square of the speed of light. An infinitesimally small amount of matter could be persuaded to release a vast amount of energy. There was the awesome prospect that you could get far more energy out of the system than you put in. The shock was how *easy* it was, and how little 'smashing' had to be done to open up the atom's heart.

Our generation sees the atomic bomb emerging as if by some inevitable force of history from the Cavendish experiments. Rutherford saw a few nuclei disintegrating, but he couldn't visualise how the accelerator might be harnessed in any practically useful way, except as a very splendid scientific tool for learning more about the atom.

On a particle-by-particle basis, yes, the energies now obtainable by the Cavendish were staggering, but basically the whole drama began and ended at a sub-microscopic scale that you could just about see through the eyepiece of a scintillation screen. There seemed no obvious way of scaling any of this up to the level of atomic power stations, let alone bombs.

The hidden weight

By the 1920s the Cavendish team knew very well that, although all atoms of the same element might behave the same way chemically, they were not necessarily identical. Francis Aston, a handsome bachelor keen on skiing, rock climbing, tennis and swimming, who could also play the piano, violin and cello, devised an ingenious machine for grading atoms by mass. His 'mass spectrometer' measured how ionised gas molecules flying through the system were deflected in their path by a magnetic field, with heavier molecules being less affected as they hurtled past than lighter ones. The molecules would finally slam into a detector screen, hitting different positions on the screen according to how far, or how little, they had been deflected. Since the missing electrons torn away in the ionising process were only an insignificant proportion of the overall mass of an atom, any difference in mass between otherwise identical atoms had to depend on something in the nucleus: something with no positive charge to it, no negative charge, and no chemical influence either. The mystery was solved in February 1932 by James Chadwick, just a few weeks before the Cockcroft–Walton accelerator won its great triumph.

Chadwick came from a poor background, and when he was accepted into Manchester University in 1908, he had to walk four miles there and back each day from his home. He was unable to participate in any of the university's social life or extra-curricular activities. He wasn't even on the right course. He was so shy that he hadn't the courage to correct the University's clerical error which sent him into a physics course instead of one for his chosen subject, mathematics. He was sufficiently beguiled to remain in physics, and in 1913 travelled to Berlin to work with Rutherford's former collaborator Hans Geiger. Then came the war, transforming Chadwick overnight from a welcome visitor to an enemy alien. He was hauled off to an internment camp, where the conditions were pretty harsh. He was allowed to study his books and make a few crude experiments from whatever bits and pieces of equipment he could beg for, but there was no question of meeting or working with other scientists, least of all Germans. He lived with four other interns in what amounted to little more than a horse stable. There was little food, and the cold winters nearly killed him. When he was at last repatriated to England, Rutherford offered him a job in Manchester. He was desperately impoverished and in poor health, but when Rutherford left Manchester to head the Cavendish, he brought Chadwick with him. In 1921 he was appointed Assistant Director of Research. By 1925 he was married, and even building a small house.

Chadwick and Rutherford already suspected that something approximately the same mass as a proton must exist inside a nucleus, even if it was fiendishly hard to identify. Some of Chadwick's tests on radioactive materials had yielded puzzling results when the radiations collided with target substances. Along with the usual alpha and beta particles, he had encountered another component, a mysterious radiation with too much momentum to be explainable as x-rays or gamma rays. 'If we suppose that the radiation consists of particles of mass very nearly equal to that of the proton, all the difficulties connected with the collisions disappear', he wrote. But the positively-charged protons and negatively-charged electrons in the atom were all accounted for. In February 1932, when Chadwick bombarded beryllium targets with alpha particles, the fragments that flew out ionised some of the hydrogen atoms suspended as gas in a sealed chamber; but he also found particles that passed right through the experiment without leaving ionised trails. Yet they had mass. He had discovered the neutron. He found, too, that neutrons had an incredible penetrative power. They could even punch deep into lead. Unimpressed by the attractions and repulsions of charged particles all around them, neutrons could slip through dense walls of atoms as though they were cobwebs. There was nothing in an atom to attract it, and nothing to repel it. Yet when neutrons resided peacefully in a stable nuclear heart, they contributed half the mass, and sometimes more. The heavier or lighter 'isotopes' of otherwise chemically indistinguishable atomic elements now had their explanation. The next great drama in the atomic story depended on the neutron. The scene was set for one of the most intimate and at the same time tragically treacherous double-acts in the history of physics.

Lise Meitner

In 1907 a delicate and painfully shy dark-eyed young Austrian student named Lise Meitner arrived in Berlin, fresh from her remarkable triumph of becoming the first woman to gain a Ph.D in physics at the University of Vienna. She hoped to study under the great Max Planck, while pursuing her interest in the new field of radioactivity. Planck was unsure about admitting a woman to his classes, but relented when Meitner turned out to be such charming company at his private musical evenings. A somewhat younger and more handsome man than Planck was also intrigued: Otto Hahn, a chemist who had trained with Ernest Rutherford at McGill in Montreal. Hahn too was working on radioactive substances. A talented experimenter and subtle careerist, he nevertheless lacked the theoretical depth and analytical intelligence he needed to propel him to greatness. Young Lise, he quickly understood, possessed these traits, yet was in obvious need of an ally in this strange new city. He proposed they should join forces. He would make the chemical separations required to obtain the purest samples of radioactive elements from the crude ores available, and Meitner would investigate their physics.

An intimate partnership developed. Or at least, it seemed to be intimate. The pair were never lovers, and they carefully observed all the social niceties so that no scandals might surround them, yet they were often heard singing duets together in their laboratory and generally sharing a deep closeness, comradely and intellectual if not actually physical. When Hahn suddenly married a young art student in 1912, Meitner told her other friends that she was unconcerned. Not many people believed her.

Hahn and Meitner shared a lab in the basement of the University of Berlin's Chemistry Institute. It was the only out-of-the-way work space that the misogynist authorities would permit Meitner to inhabit, and it took all of Hahn's charm to achieve even that. Even so, Meitner made an impression and quickly rose to academic respectability under Planck's increasingly enthusiastic patronage. This didn't go unnoticed at the Chemistry Institute, which moved swiftly to improve Meitner's working conditions. By the beginning of the First World War in 1914 she and Hahn were making excellent progress in the newly-founded 'Hahn–Meitner Laboratorium'.

Both survived the war years, even though Hahn saw active service and Meitner drove ambulances across the battlefields. Working together on and off throughout the 1920s and early 1930s, they made wide-ranging contributions to the study of atoms, including the discovery and purification of hitherto unknown elements: gaps in the heavy end of the periodic table. Then they raced against rival teams in Germany, Paris and England to identify a new and extremely penetrating kind of radiation, stimulated when alpha particles bombarded light elements such as beryllium and lithium. Finally, in 1932 Rutherford's former pupil James Chadwick worked out what this radiation really was: the neutron. In time, Meitner would understand the neutron's true power. First she had to adjust to some terrifying changes in Germany.

In Hahn's way

Meitner's Jewish origins threatened her prospects after Hitler came to power in 1933, but since she had been born an Austrian citizen, she thought she might be safe. She was desperately unwilling to abandon all the work she had done in Berlin, and imagined that the hard-won status she had achieved among her fellow academics could protect her. When Germany annexed Austria in 1938, Meitner became a German citizen by default. Her rivals within the Chemistry Institute could now whisper a little more loudly: 'The Jewess endangers our Institute.' Alarmed, she talked the matter through with her old friend and colleague Hahn, who promised to try to help. He then requested a private meeting with the chief administrators of the Institute. And he agreed with his superiors that they should get rid of Meitner.

The science writer David Bodanis identifies one of Hahn's essential characteristics: 'To say that people have been charming, as he had been all his life, is simply to say that they've developed a reflex to do what will put the individuals around them at ease. It says nothing about their having a moral compass deeper than that.' Hahn had failed Meitner to the utmost degree that one person can betray another. Shocked and terrified, she made her way to Sweden, slipping past the German border guards with her old and recently expired Austrian passport. Fortunately they did not choose to ask too many questions, and Meitner escaped. She could very easily have ended up in a concentration camp, yet her attitude towards Hahn was remarkably forgiving, and she somehow persuaded herself that he had acted out of necessity.

Given the terrifying pressure on all German scientists to comply with Nazi ideology, perhaps Hahn's move to evict her was unavoidable. The strange collaborators continued writing to each other, and even held a meeting in Copenhagen. Meitner's career prospects in Sweden were limited, and she had no access to the kind of laboratory she needed; so she arranged to write detailed notes to Hahn, while he conducted practical experiments back in Berlin, aided by his and Meitner's former assistant Fritz Strassman.

The two men discovered that bombarding uranium with slowed neutron beams produced apparently new and lighter subatomic components, although they could not immediately identify them. It all seemed too much at variance with what they thought they might find – uranium atoms transmuted into heavier isotopes by the insertion of extra neutrons. Hahn made a difficult chemical analysis of the extremely tiny products created by the experiments, and concluded that one of them might be barium. None of the calculations made any sense. It was just about conceivable for elements to be transmuted from one to another so long as the newly formed products had a similar atomic weight to the first one, but barium was half as light as uranium. It seemed too much of an alchemical leap. Puzzled and excited in equal measure, Hahn wrote to Meitner a few days before Christmas 1938, asking what all this might mean, and promising her that 'if there is anything you could propose that we publish, then it would still be work by the three of us. You will do us a good deed if you can find a way out of this.'

It dawned on Meitner that something extraordinary had happened. Hahn and Strassman had shattered massive uranium atoms using even slower-moving particles than Walton and Cockcroft had employed to break lightweight lithium nuclei. Slow neutrons insinuated themselves into the uranium nuclei and upset the balance between the protons' powerful desire to repel each other and the nucleus's equally powerful grip on them, via a binding force called the 'strong nuclear interaction'. Even before an additional neutron was added, a uranium atom was already a precarious thing. The rarest isotope, uranium-235, was overloaded with 92 protons and 143 neutrons. The extra neutron's insertion might have been unexpectedly delicate, but the subsequent flying-apart of the uranium nuclei was dramatically less so. But why did the gentle insertion of a neutron deliver such a spectacularly violent disruption? Where did the immense energy come from, when the invading neutron had been moving so sluggishly? Meitner recalled a meeting with Albert Einstein at a conference in Salzburg back in 1909, and recalled his 'overwhelming and surprising' prediction that matter was essentially a super-compressed form of energy – energy that could be released, so that it would appear startlingly greater than you might ordinarily expect from the tiny speck of matter that produced it. (Likewise, the Cavendish team had been startled by the violence of the alpha particles that flew out of shattered lithium nuclei in Cockcroft and Walton's accelerator experiments.)

Meitner's nephew, the young physicist Otto Frisch, another exiled German, had found refuge at Bohr's Institute in Copenhagen. Now he rendezvoused with his aunt at a quiet holiday cottage on the Swedish coast, lent to her by sympathetic friends; and here, surrounded by deceptively calm natural beauty, they discussed the incredible implications of breaking atoms apart. 'The uranium nucleus, we found, might indeed resemble a very wobbly, unstable drop of water, ready to divide itself at the slightest provocation, such as the impact of a single neutron,' Frisch recorded. When a uranium atom was shattered, its exploded fragments apparently consisted of two new nuclei, roughly equal in size. The energy for this great burst did not have to be supplied from outside the experiment. It was already contained *in* the atom. What was more, there would still be two or three neutrons left over from this split, and these would be travelling slowly enough (liberated, rather than hurled) so that they could slip into the nuclei of adjacent uranium atoms, upsetting their internal balances, and so forth, in a chain reaction that Frisch likened to the splitting or *fission* of breeding bacteria.

As he later recalled, when he and Meitner accounted for all the protons and neutrons in the uranium atom, something was still missing. 'My aunt worked out that the two nuclei formed by the division would be lighter than the original uranium nucleus, by about one-fifth the mass of a proton. Now whenever mass disappears, energy is created, according to Einstein's formula $E = mc^2$.' One-fifth of the mass of a proton seemed a very small amount of matter. On the other hand, the square of the speed of light is an unimaginably huge number. Atomic fission could liberate vast amounts of energy … Three days into the New Year, Meitner sent a telegram to Hahn: 'I am fairly certain now that you really have a splitting, and I consider it a wonderful result for which I congratulate you and Strassman very warmly. You now have a wide and beautiful field of work ahead of you.'

Frisch returned immediately to Copenhagen and told Bohr the news. Bohr was in a mad rush packing for a trip to America. He had a great talent for nearly missing trains and ships, and he listened excitedly to Frisch's news about fission while still packing his bags. 'What idiots we have all been! Oh but this is wonderful! This is just as it must be!' What with all the business of checking his travel documents and so on, he paid slightly less attention than he should have done to one small detail. Meitner and Frisch hadn't published their ideas yet, so the news was supposed to be kept secret. Once aboard his ship, Bohr told an associate the news, and in his excitement he forgot to enforce the secrecy. The news hit America, where it created an absolute sensation, much to Bohr's embarrassment. Journalists were clamouring for more information, and it was becoming impossible for him not to talk about it. He wrote a guilt-stricken letter to his wife Margarethe: 'I was immediately frightened, as I had promised Frisch I would wait until Hahn's paper appeared, and his own was sent off.' He also cabled Frisch in Copenhagen, urging him to publish his and Meitner's calculations as soon as possible.

On 26 January 1939, the Fifth Washington Conference on Theoretical Physics was held at George Washington University – the prior engagement which had caused Bohr to head for America in the first place. Just in time, he learned that Otto Hahn had published his own paper on 19 January, although he had phrased his laboratory results somewhat cautiously. Some alert journalists had obtained a copy of the paper, which they quickly showed to Bohr. At last he was free to let the cat out of the bag, although his announcement at the Conference was so garbled and hesitant, very few people understood the significance of what he was trying to communicate anyway. Enrico Fermi, a physicist recently exiled from fascist Italy, took up the slack and made a better job of it. Fission itself then spread like … fission. Within the week, laboratories at Berkeley, Baltimore and Columbia announced that they too had split the uranium atom. Now that everyone knew what to do, the experiments took only a short time to prepare.

In the prevailing climate Hahn could not, of course, acknowledge Meitner's contributions without jeopardising his safety. What is less understandable is that he spent two decades *after* the fall of the Nazi regime claiming that the major breakthrough had been his alone. In 1944 he collected the Nobel Prize 'for his discovery of the fission of heavy nuclei'. There was no mention of Meitner in that citation either. She understood his reluctance to name her in his original paper, but was not so sanguine about his continued failure to credit her once the war was safely over. Years later she wrote to a friend: 'I found it quite painful that in his interviews he did not say one word about me, or anything about our years of work together. His motivation is somewhat complicated. I am part of the suppressed past.' Arguing about credit began to seem less important to Meitner after 1939, as the greater and more terrible potential of her discovery came close to realisation. It looked as if the fission of a single uranium atom could liberate enough energy to make an entire grain of sand jump in the air. A small lump of uranium might shake all the grains of sand that make up a city ... Meitner began desperately telling anyone who'd listen: 'I will have nothing to do with a bomb!'

PART FIVE

Blast Radius

How could people of conscience create weapons of mass destruction? We live in an age of localised terrorism, and sometimes we forget that the atom bomb was once the dominant threat to all life on earth. We also forget that the bomb is still here. It hasn't gone away.

The most famous Jewish refugee scientist, of course, was Albert Einstein. He was already thoroughly accustomed to restless dislocations. In 1909 he had accepted a professorship in Zürich, then in 1911 he was appointed as Professor of Theoretical Physics at Prague, returning to Zürich in the following year to fill a similar post. In 1914 he became Director of the Kaiser Wilhelm Physical Institute, and also a professor at the University of Berlin, the very same organisations that had sponsored so much of Meitner and Hahn's work. He took up German citizenship in 1914 and remained in Berlin until 1933, when it became clear that his ancestry put him at risk. It seems unimaginable today that someone of his stature could ever have been threatened, but the Nazi regime's tame academics condemned his world-famous theories as 'Jewish science'. He renounced his German citizenship (for the second time) and emigrated to America to take the position of Professor of Theoretical Physics at the Institute for Advanced Studies at Princeton, New Jersey. Here he adopted his new guise as a much-respected yet faintly out-of-date wise elder statesman of science. This is the period of the genial middle-aged man with the crazy shock of white hair so familiar to us from countless photographs of Einstein 'the genius'. He continued to fight his losing battle against the Young Turks of quantum mechanics, but by now his energies were depleted by other calls on his time, his emotions and his moral sensibilities.

In mid-July 1939 he was enjoying a sailing holiday on the northern tip of Long Island, New York, when he was surprised by two visitors, a pair of Hungarian–Jewish refugee physicists, Leo Szilard and Eugene Wigner. Their car pulled up to Einstein's holiday cabin early in the morning, and the great man answered the door to them half-dressed. Somewhat bemused, he showed them to his cabin's large, screened-in porch, and listened to what they had to say. Szilard warned that the occupying German forces in Czechoslovakia had blocked all international exports of uranium from the Joachimsthal mines. It seemed reasonable to conclude that they must be trying to build a nuclear device, and almost certainly they would wish to develop a uranium bomb. Szilard asked if Einstein could warn the Belgian Queen Mother, whom he knew as a friend, to prevent the large stockpile of uranium ore in the Belgian Congo from falling into Nazi hands. Einstein agreed to the idea, and the three men decided that his name might carry some weight, if only they could find a suitably senior audience for their views. A few days later, Szilard discussed the problem with Alexander Sachs, an economics adviser to President Roosevelt. Sachs urged that Einstein should write directly to the President, and promised that he'd deliver the letter personally. Einstein now had a direct route into the Oval Office.

F.D. Roosevelt, President of the United States,
White House, Washington D.C.
August 2, 1939

Sir:

Some recent work ... leads me to expect that the element uranium may be turned into a new and important source of energy in the near future. Certain aspects of the situation which has arisen seem to call for watchfulness and, if necessary, quick action on the part of the Administration ... It may become possible to set up a nuclear chain reaction in a large mass of uranium, by which vast amounts of power ... would be generated ... This new phenomenon would also lead to the construction of bombs, and it is conceivable – though much less certain – that extremely powerful bombs of a new type may thus be constructed.

Our popular image of Einstein as a benign guru and kindly man of peace sometimes blinds us to a more complex truth about his attitude to the bomb. In his letter to Roosevelt he was *not* arguing that it should remain unbuilt. He urged the President to 'secure a source of uranium for the United States ... and speed up research'. He also insisted that small-scale university experiments were no longer sufficient. America would have to create a large-scale industrial response. Knowing perfectly well the risks of encouraging America even to consider this technology, Einstein was swayed by his premonition of a far greater evil. If the Nazi regime managed to create an atomic bomb unopposed, it would surely use it. For ten frustrating weeks, the White House made no reply. Roosevelt was preoccupied by military events in Europe, but at last he wrote to Einstein on 12 October that the American government wished to investigate the uranium question at an official level. It would be yet another two years before anything much more tangible happened, but the scene was set for the 'Manhattan Project', and the coming of the atom bomb.

Certainly one other man in America had already understood the darker possibilities of fission. Just a day after news of Meitner and Frisch's conclusions had leaked across the Atlantic in January 1939, Robert J. Oppenheimer had a sketch of a bomb on his blackboard at the University of California, Berkeley.

Now teach me something

Like so many physicists, Enrico Fermi had shown an extraordinary talent for mathematics at a very young age. He was born in Rome in 1901, the youngest of three children. His mother, Ida, was a schoolteacher, and his father Alberto a railway administrator. These were respectable jobs, although not particularly well paid. Roman summers were easy enough to survive, but in the winter, when fuel ran short, Enrico would study his books while sitting on his hands to keep them warm. He'd lean down and turn over the pages with his tongue. It was a very happy family until 1915, when Enrico's older brother and favourite playmate Giulio died unexpectedly in hospital during what should have been a simple throat operation to treat an abscess. The Fermi house was plunged into mourning. Ida had a complete breakdown, while Enrico took lonely comfort in his books, beginning an intense study of mathematics and physics.

An alert schoolteacher persuaded the Fermis to send Enrico for an interview to the Scuola Normale in Pisa, formerly a teacher training institute, now a small and far from 'normale' school accommodating just 40 of Italy's best and brightest youngsters. Enrico stood out even from this tiny élite. As a test prior to admission, he wrote an essay on the characteristics of sound waves, diligently including all the appropriate equations and their correct solutions. The Scuola Normale's physics teacher was impressed, and, unlike many of his position and rank, gladly reconciled himself to the idea that his pupil might be brighter than he was. He'd sit the boy down in his office, and with an amiable smile he'd say: 'Now teach me something.' And young Fermi would oblige, for he loved to explain science as well as to discover it.

Fermi was never short of such patrons, and his career prospered. The Italian scientific establishment was acutely aware of its lack of talent in physics. In 1926 Fermi easily won the competition to take up the new Chair of Theoretical Physics at the University of Rome. At just 25 years of age, he had achieved one of the most senior positions any university has to offer. Over the next decade, at his beloved laboratory at the Physics Institute on Via Panisperna, Fermi worked out *why* neutron rays could penetrate other substances much more deeply than alpha or beta particles. It was because they swept right past the usual blockages of positive charge from protons or negative charge from electrons. They could be used, in fact, to probe atoms – but only if they could be slowed down so that they wouldn't simply push right through them. In 1934 Fermi produced a stream of 'slow neutrons', impeding their progress with water taken from his backyard pond and, later, a 6-centimetre-thick shield of paraffin, which, he discovered by trial and error, did the trick even better. Lightweight hydrogen atoms in the paraffin slowed down the neutrons without actually blocking them. Puzzled by the emitted products of these experiments, Fermi announced that he had transformed uranium atoms into new and heavier isotopes whose radioactivity decayed after just a few seconds into lighter nuclei. He wasn't absolutely certain about the results he'd achieved. Without quite realising it, he had in fact split some of his uranium atoms.

Rutherford sent him a congratulatory letter, while researchers around the world rushed to explore the investigative possibilities, whatever they might turn out to be. None of them truly understood what they were about to unleash. At the time, the great prize seemed merely to be that these slow-moving neutrons could be gently inserted into uranium nuclei, adding to their mass and transmuting them into heavier 'transuranic' elements. It seemed nothing more than a Nobel Prize-winning intellectual game, a harmless tinkering with the atomic numbers in the periodic table of elements. In 1933 Ernest Rutherford, in his early sixties by now but still boundlessly energetic, made a radio broadcast for the BBC outlining some of the latest discoveries: 'My listeners may quite naturally ask why these experiments should excite such interest in the scientific world. It is not that the experimenter is searching for a new source of power,' he asserted confidently. It was all about the 'search for the deepest secrets of Nature'.

Up until 1937, Fermi had no particular grievance with the Mussolini regime in Italy, so long as it left him alone. Sometimes he could reach his laboratory only by negotiating his way past armed patrols on the streets of Rome. Dressed shabbily and driving a battered car, he liked to amuse himself by telling the guards that he was the chauffeur for a very important man, a certain Professor Fermi, so they'd better let him through. When Mussolini made an alliance with Hitler and agreed also to abide by his racist policies, Fermi's Jewish-born wife Laura was endangered. In 1938 Niels Bohr broke the usual academic protocols and told Fermi, in confidence, that he was about to be awarded a Nobel Prize for his work with slow neutrons. Bohr didn't need to spell out the implications. The Fermis travelled to Stockholm to collect the prize and then headed straight for America, where the Italian scientist had little difficulty integrating himself with the Americans and their Manhattan Project.

In the autumn of 1942, Fermi commandeered an old squash court under the spectator stands of Stagg Field, Chicago University's antiquated football stadium, recently abandoned in favour of a smarter new one across the campus. Curious passers-by were told that his peculiar new lab was working on problems in metallurgy. None of the thousands of people milling through the campus knew that the ramshackle device taking shape behind the locked doors of the squash court had the capacity to kill them all.

The pile

In the centre of the 30- by 60-foot room, shrouded by huge sheets of grey balloon cloth, lay a pile of black bricks and timbers, square at the bottom, and with a slightly flattened ziggurat on top, rather like the uppermost steps of a pyramid. Deeper inside the 'pile', as it was known, layers of graphite blocks were interleaved with layers containing blocks of uranium. It was a crude-looking machine, and certainly didn't look like value for the huge sums of cash that had been spent on it. Fermi's team liked to joke that: 'If people could see what we're doing with a million-and-a-half of their dollars, they'd think we're crazy. If they knew why we are doing it, they'd *know* we are.' Yes, it was scary, and the original scheme had called for the pile to be built in a secluded forest preserve some twenty miles south-west of Chicago. When union construction workers went on strike, the building designed to house the pile would not take shape as fast as the war effort required. Drastic measures were called for, and the experiment was transferred to the Stagg Field site. It seemed insanely hazardous, but Fermi assured his financial sponsor Arthur Compton, Chairman of the 'National Academy of Sciences Committee to Evaluate the Use of Atomic Energy in War', that he could keep the pile under control. His grid of uranium blocks was interspersed at regular intervals by horizontal shafts, into which 'control rods' of cadmium-plated graphite could be dropped to block any free-flying neutrons and damp the reactions. It was yet another of Fermi's key discoveries that graphite, the form of carbon that we find in pencils, was a better and more easily handleable neutron moderator than anything else he'd experimented with. The cadmium plating lent the friable graphite control rods some rigidity. From now on the Manhattan Project, and indeed almost all subsequent nuclear reactor research, would rely on it.

Compton was convinced by the technical arguments, yet he faced a tricky political decision. Quite apart from his role

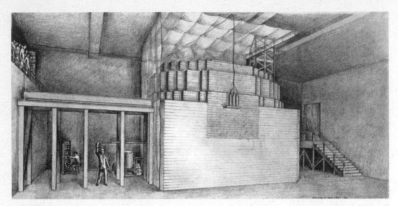

Figure 3. A sketch of Enrico Fermi's uranium pile under the Stagg Field stadium in Chicago. Very few photographs of the world's first nuclear reactor exist. Stray radiation leaking from the pile was strong enough to fog photographic film. (Archival Photofiles (apf2-00503), Special Collections Research Center, University of Chicago Library)

on the secret National Academy committee, he was also an ordinarily accountable physics professor at the University of Chicago ... Briefly he considered running Fermi's plan past his University superior, Robert Hutchins. But Hutchins, as President of the University, would have been irresponsible to say anything other than 'no' to such a dangerous scheme. 'During war, one takes risks,' Compton later said. Without giving Hutchins the full facts, he gave Fermi the green light to build his pile in the heart of the University.

Fermi was reduced to using a less than perfect variety of uranium ore containing only a 7 per cent proportion of the particular isotope he needed. Therefore he needed a lot of ore, and an equivalently large amount of graphite to house and control it. But the prize on offer from the Stagg Field experiment was something far more compact and efficient. Glen Seaborg, one of Oppenheimer's co-workers at Berkeley, had recently discovered that a new element, more suited for a chain reaction than even the best and purest isotopes of uranium, could be created as a by-product of carefully controlled uranium fission, in which some neutrons could be induced to stay within the uranium nucleus rather than splitting it. This new element came to be known as 'plutonium' in honour of Pluto, at that time the last known planet in the solar system. Plutonium occupied an ambiguous slot in the periodic table at a point where naturally occurring substances stopped and the eerie products of human alchemy began. After all, it was just this kind of super-heavy 'transuranic' element that the nuclear researchers had been looking for when they first stumbled across atomic fission ...

After just six weeks of construction work, the Stagg Field pile was activated on 2 December 1942. The cold, icy weather that day belied the ferocious energies now being unleashed. Fermi and his colleagues listened, almost in shock, as the surrounding neutron counters clicked with increasing rapidity until they were off the scale and the microphones generated a roar of white noise. Left to its own devices, the pile would have generated so much energy that it would have melted, killing everyone in the lab and poisoning a giant swathe of the University district with intense radioactivity. After four minutes and 30 seconds, Fermi ordered the control rods to be reinserted, and the neutron counters' fierce clicking subsided.

There was no announcement, and no fanfare in the press about this history-making moment. The nature and purpose of this, the world's first sustained nuclear chain reaction, was so secret that Fermi hadn't even told his wife what he was doing. The way was now clear to manufacture worthwhile quantities of plutonium inside larger industrial-scale uranium reactors. Then the relatively tiny amounts of plutonium would have to be extracted from the surrounding uranium dross and purified using a combination of chemical and electromechanical processes. The ultimate product of this multi-billion-dollar endeavour would be a lump of plutonium small enough to fit inside a bomb. By early 1943, two huge nuclear manufacturing facilities were taking shape at sites located conveniently near powerful hydroelectric plants: one at Oak Ridge in Tennessee, biased towards work with uranium, and a second at Hanford in eastern

Washington State, whose focus was on plutonium. These were colossal facilities. At its height, the Manhattan Project accounted for some 140,000 personnel and absorbed $2 billion in 1940s money (equivalent to $20 billion today).

The third centre, the very heart of the Manhattan Project, had no need for such raw natural power. Its special requirement was for absolute secrecy and remoteness. 'Project Y' was founded a hundred miles into the desert north of Albuquerque, in a sleepy New Mexican desert town called Los Alamos. Here, a thoughtful and somewhat mournful scientist named Robert Oppenheimer convened some of the finest theoretical physicists the world could offer – including many whom Germany had carelessly discarded.

The 'nim-nim-nim' boys

Born into an affluent Jewish family and raised in a cultured home decorated with fine paintings and elegant furniture, Robert Oppenheimer benefited also from an unusually broad education. As a youngster he attended the Ethical Culture Fieldston School in New York, where students were encouraged to learn not just the conventional facts and figures required to pass examinations, but also the moral and ethical contexts of life. He learned eight languages, became fascinated by classical subjects such as Latin and Greek, and made himself thoroughly aware of many literatures and philosophies, both modern and ancient. He studied chemistry at Harvard, then crossed the Atlantic to work at Cambridge University under J.J. Thomson at the Cavendish, where he found his day-to-day practical work, preparing thin films of beryllium for electron bombardment, deadly dull. He escaped to Germany and the University of Göttingen, where he met and worked regularly with Bohr, Heisenberg, Pauli and others, travelling widely across Europe and steeping himself in the emerging new theories of quantum mechanics. By his 23rd birthday, his reputation as a talented theoretician was already growing.

In 1927 he returned to America, where he was offered dozens of teaching posts by enthusiastic faculty chiefs from around the country, all keen to absorb the latest news from Europe. Faced by such a luxury of choice, he decided to spend his spring semesters at the California Institute of Technology (Caltech) in Pasadena, and the rest of his time at Berkeley, where he became the key man in America's first truly serious efforts to make its own advances in theoretical physics. He also took a considerable interest in left-wing politics, especially when the fascist movements in Europe gained such terrible strength in the mid-1930s. He was keen, too, on Eastern mysticism, and liked to quote from the *Bhagavad-Gita* in its original Sanskrit. His students fondly emulated his verbal tics and cultural mannerisms. Eccentric and sometimes bohemian, a tall, restless, chain-smoking stick figure of a man with long, flailing arms, he was a distinctive and popular character, although his soft, halting delivery and frequent pauses for thought frustrated some of his listeners. He would stutter in curious ways, which his students fondly tried to pin down and mimic. When someone suggested to Wolfgang Pauli that no useful physics was being done in America, he replied: 'You mean, you haven't heard of Oppenheimer and his "nim-nim-nim" boys?'

For all this gentle stammering, and the deep sense of cultural morality that Oppenheimer had absorbed as a young student in New York, he could be quite superior and hurtfully sarcastic in his personal relationships. These traits stemmed from his insecurities. He was not a contented person at ease with himself. It was a strange quirk of history, then, that he should become one of the kindest and most supportive leaders of a project whose purpose was to make possible the delivery of death and suffering on an unprecedented scale: the Manhattan Project, America's bid to build the first atomic bomb. By November 1941 the Roosevelt White House had absorbed the warnings about Germany's bomb programme and had called for a response. A month later, Japan attacked Pearl Harbor, and America went to war. The Manhattan Project took shape under the hard-driving leadership of an Army general, Leslie Groves.

This vast enterprise became one of the greatest industrial achievements in history. How is it that our darkest engineerings so often possess a dreadful grandeur that cannot help but fascinate us? We love the elegant lines of the Spitfire and admire the swift efficiencies of the Messerschmitt. We mourn the loss of the beautiful battleship HMS *Hood* and send submarines into the deep realms of the Atlantic to study the superbly engineered wreck of the *Bismarck*. We flock in our tens of thousands to air shows in which silver-winged killing machines send sixteen-cylinder rumblings of guilty glee through our rib cages. We celebrate how *brilliant* were those men and women who made 'the bomb'.

Uncertainty about Heisenberg

By the summer of 1937 Werner Heisenberg was renowned as one of the world's greatest physicists, almost as great as Einstein. He had recently wooed and won a beautiful wife, Elisabeth Schumacher, and had returned to the old family apartment in Hamburg after a wonderful honeymoon. He was looking forward to a new and prestigious appointment in Munich, at the very same university where he had earned his Ph.D. He had few qualms about pursuing his researches under the Hitler regime, even if he did not specifically condone many of its values. Much to his angry surprise, a superannuated rival professor and virulent anti-Semite, Johannes Stark, persuaded the SS-sponsored magazine *Das Schwarze Korps* ('The Black Corps') to run an article questioning his patriotism. Was it not true that he had worked for many years alongside Jewish scientists? Did he not believe in the 'Jewish science' of Einstein?

Heisenberg took a risk, writing to SS chief Heinrich Himmler to ask that these charges be formally investigated so that he could clear his name – whatever that meant. An unsettling sojourn in the basement rooms of SS headquarters at Prinz Albert Strasse in Berlin shook his confidence. He began to think he might be in serious trouble. Fortunately his mother was on good terms with Himmler's, and the two matriarchs met to discuss the problem. Frau Himmler wasn't sure if it was appropriate for her to intervene in her son's important affairs, so Frau Heisenberg went straight for the maternal weak spot: 'We mothers know nothing about politics, neither your son's or mine. But we know that we have to care for our boys.' It worked. Himmler wrote personally to Heisenberg assuring him that the attacks against his character would cease. Just as suddenly and capriciously as they had been endangered, Heisenberg's prospects in his homeland were now greatly enhanced by Himmler's protection.

In 1939 Heisenberg travelled to the United States, where he lectured at Ann Arbor and in Chicago, and caught up with many old friends. In his memoirs, written some two decades after the war, he suggested that 'if I was to contribute to Germany's reconstruction after the collapse, I would badly need their help.' Perhaps he'd recognised as early as 1939 that Germany could never succeed in its grand ambitions for European dominance. It's more likely that he adjusted some of his memories after the fact. In Chicago, Enrico Fermi tried to persuade him how much happier he'd be if he stayed in America. He chose to return to Germany, and specifically to Berlin, where the Kaiser Wilhelm Institute for Physics, now known to insiders as the Uranium Club, became the nexus for German nuclear research. Its first director, Peter Debye, travelled to Switzerland on a lecture tour and decided not to return, and Heisenberg emerged as the Uranium Club's de facto star. By 1942 there could be no uncertainty about his status as the head of Nazi Germany's uranium fission project. The mere fact that he was in charge gave urgent impetus to America's bomb-makers. This much is known. Historians have spent the last half-century debating how close Heisenberg and his team came to building their own bomb – or how hard they tried.

Heisenberg's visit to occupied Copenhagen in the autumn of 1941 shows how history itself is subject to its own brand of 'uncertainty principle', in which the perspective of the observer determines the outcome of an encounter. After the war Heisenberg insisted that his intention had been only to seek advice from his reliable old ally Niels Bohr. 'Does one as a physicist have the moral right to work on the practical exploitation of atomic energy?' In an interview with the BBC in March 1963 he remembered telling Bohr: 'It requires such an enormous industrial effort that probably the bombs won't be ready until the end of the war.' The project would cripple Germany's military strength, and he assumed that any similar project on the Allied side would be just as vast, and he wondered – or so he claimed – if it might be better if everyone agreed that 'it's no use to build atomic bombs'. Bohr's memory of events was subtly yet crucially different. In his version, Heisenberg insisted that 'the war, if it lasted sufficiently long, would be decided with atomic weapons'.

Bohr had good reason to feel sensitive. In 1933, as soon as Hitler rose to power, he travelled to Germany and made it known to dozens of scientists that he would welcome them at his Institute in Copenhagen. He had to tread carefully. What he was actually doing was laying down emergency escape routes for those endangered by Nazi racism. The Danish High Court also moved with admirable decisiveness, establishing a Committee for the Support of Refugee Intellectuals. Bohr and his advisors made a compelling argument. Human decency demanded that endangered scientists should be offered a route out of Germany, while cold-hearted pragmatism dictated that the Nazis alone could not be allowed scientific supremacy in the coming dark times, so any talented people that the Germans wilfully and stupidly threw away should be welcomed.

Inquiries from post-war historians about that 1941 confrontation with Heisenberg left Bohr deeply upset. In one draft of a letter still held at his Copenhagen Institute, which he dictated in a fury in March 1961 but never actually sent, he wrote: 'Dear Heisenberg, I am greatly amazed to see how much your memory has deceived you. Personally, I remember every word of our conversations, which took place on a background of extreme sorrow and tension for us here in Denmark.' He wanted to remind Heisenberg of his 'definite conviction that Germany would win and that it was therefore quite foolish for us to maintain the hope of a different outcome of the war ... I also remember quite clearly our conversation in my room at the Institute, where in vague terms you spoke in a manner that could only give me the firm impression that, under your leadership, everything was being done in Germany to develop atomic weapons.' In 1967 Heisenberg told *Der Spiegel* magazine a different story. 'We in Germany suddenly saw a path to an atomic bomb that, at least in principle, was technically feasible and this situation itself was quite horrific to us. We had unlimited trust in Bohr as one of the leading atomic physicists. His human advice was bound to matter a lot to us. Unfortunately though, in this conversation we were not really able to communicate.'

The best we can surmise about Heisenberg is that he did not believe he could build a bomb with the resources available, or in the short time that the Nazis would have demanded of him if he'd promised to deliver one. After the war he claimed that he avoided asking for large sums of money because he wanted to slow the bomb's development. It's much more likely that he knew his failure to make one would put him in danger, so he promised his terrifying employers only what he thought he could achieve. He always did believe he could create a uranium reactor. If Germany had succeeded in its war aims, and if Heisenberg had felt confident enough to demand more time and money, no one has any reason to believe that he would have shied away from making a bomb, or from seeing it used. In 1942 he tinkered with a strange paper on philosophy, always one of his pet enthusiasms. One phrase is particularly telling. 'We should transfer to the next generation that which still seems beautiful to us, build up that which is destroyed, and have faith in other people above the noise and passions. And then we should wait to see what happens. Reality is transformed by itself without our influence.' He seemed to be a fatalist who did not believe that he or any other individual could reshape the overwhelming forces of history. There were gigantic 'movements of thought' sweeping across Europe. The individual, he believed, 'can contribute nothing to this, other than to prepare himself internally for the changes that will occur anyway'.

On the other hand, he did work energetically at his own small scale to safeguard his reputation, no matter what the chaos all around him. His race to try to finish a reactor in the last years of the war, even as Germany crumbled, may have been his attempt to make himself an attractive prize for whichever regime came next, whether German, Russian or Anglo-American. Heisenberg was extremely shocked in 1945 when he learned how far ahead of him the American atomic team had always been. He set about rewriting his history, and showed little of Oppenheimer's desire for harsh self-analysis and sorrowful introspection. Thomas Powers, author of a major study of Heisenberg's life, elegantly finds the link between his science and the ambiguities in his personal history: 'Questions of motive and intention cannot be established more clearly than he was willing to state them.'

The belly of a Mosquito

Bohr was a very different character, one who believed that individual actions counted for everything in a world plunged into darkness. He had worked in league with the Danish authorities to smuggle almost all of Denmark's 7,000-strong Jewish population into Sweden before the Nazis could capture them. A marked man, he nevertheless accepted his government's judgement that if he left Denmark it would set a demoralising example to his fellow citizens. He stayed as long as he could, escaping only in October 1943 when the danger became intolerable. He and his family arrived safely in Sweden, after which he almost immediately set out on a dramatic solo journey to England, where the government was anxious to hear his advice about atomic fission, and to learn whatever he might know about Heisenberg's capabilities. He boarded an RAF Mosquito fighter-bomber (unarmed so as not to endanger Sweden's neutrality) but there was no room for him in the two-man cockpit, so he was bundled up in the bomb bay with a parachute harness and a handful of flares. He was instructed that if the plane was hit by enemy fire he should parachute into the water and light his flares. Rescuers would be looking out for him. What they didn't tell him was how important it was to keep his oxygen mask fitted snugly against his mouth and nostrils. He spent most of the journey to England unconscious.

Once there, he mediated between British atomic scientists and the American bomb team. Or rather, he met face to face with people he'd already been talking to via covert telegrams. In July 1941, a secret British advisory group code-named the MAUD Committee studied the possibility of nuclear weapons and concluded that just one kilogram of purified uranium-235 might be enough to produce an explosion. And there would be no need for a huge mass of graphite moderator inside a bomb, because the 235 isotope should fission with *fast* neutrons. This all meant that the entire device could be made to fit inside the bomb bay of an aircraft. When this news was conveyed to the Manhattan team, they were naturally very interested. Bohr and Otto Frisch were among the theorists whose contributions had made the MAUD report such compelling reading, although the acknowledged authorship belonged to James Chadwick, John Cockcroft and other British scientists. Beleaguered Britain could not afford its own bomb, but MAUD had certainly proved its mettle to the Americans, whose initial calculations about uranium had been orders of magnitude less encouraging, at least until the plutonium option became apparent. Unfortunately, given that the British intelligence agencies at that time were comprehensively infiltrated by Soviet double agents, MAUD also alerted Russia that it was time to start thinking along the same lines.

The name 'MAUD' was not an acronym, more the commemoration of a misunderstanding. At one point Bohr had sent Frisch a telegram, carefully worded as ever, in which he oh-so-casually asked Frisch to send his regards 'to Cockcroft and Maud Ray Kent'. British intelligence teams had no problem working out who Cockcroft was, but 'Maud' was a mystery. Was Bohr trying to communicate some strange atomic concept? No, he really meant Frisch to send his regards to Maud Ray Kent, a former governess to the Bohr's children, who had since moved to England.

James Chadwick, Rutherford's right-hand man and discoverer of the neutron, led a British team who joined the Manhattan Project. Bohr then made several visits to Los Alamos, where he was an invaluable comfort to men torn between the thrill of building something that could shine with the light of a thousand suns and the horror of killing tens of thousands of people. Oppenheimer said of him: 'He made the enterprise, which often seemed so macabre, seem hopeful. He spoke with contempt of Hitler, who had hoped to enslave Europe. He said nothing like that would ever happen again … All this was something we very much wanted to believe.'

In Bohr's absence, his precious Institute at Copenhagen was temporarily occupied by Gestapo troops, although none seemed to have any idea what to do once they got into the building. Heisenberg made another mysterious visit, supposedly to try to protect the Institute against careless damage. After all, he had every reason to be fond of the place. Nevertheless, the Danish underground knew to be wary of him, and made its plans accordingly. Underground workers crept through the sewers and planted dynamite beneath the building. A flurry of secret messages went back and forth in a last-minute effort to contact Bohr and see if this was what he wanted. He managed to intervene just in time to save the Institute. Shortly afterwards, the Gestapo moved out as swiftly and pointlessly as they'd moved in. The building was undamaged, but the same could not be said of the close companionship that had once existed between Bohr and Heisenberg.

After all the fuss and paranoia, Heisenberg may not actually have been the great bomb-maker the Allies feared him to be. He made some profound errors, especially in the way he shaped his uranium during his early fission experiments. He never achieved the breakthrough he (presumably) wanted. A brilliant theorist, he could have done with the services of a good practical engineer. Anyone who had spent their working lives around high-temperature gases and combustion processes would have understood the problem. Just like any other great onrush of energy, the discharges of the neutrons had to be steered, gathered and concentrated in a more efficient way. Fermi's uranium pile under the old football field at Chicago University was built with that in mind. At the Manhattan Project, Oppenheimer knew to employ experts in conventional explosives who understood how to shape the force of released energies. Their calculations were vital in the making of the bomb. Heisenberg was not that kind of an engineer, and saw too late that his reactor would work better if the uranium was cut into blocks and arranged around a focused core, rather than laid out in his less imaginative array of parallel sheets.

His other problem, perhaps the most compelling, was that his neutrons were almost always too fast. Unlike Fermi and his team in Chicago, he had decided quite incorrectly that graphite, a fairly simple substance to prepare and handle, was no good as a moderator. If only the Nazis hadn't discarded so many brilliant scientists, this misjudgement could have been quickly corrected. Heisenberg turned instead to paraffin, while waiting for sufficient quantities of a special kind of water that was available only from one place: Norway.

The heroes of Telemark

Hydrogen is at once the most abundant element in the universe and the simplest. It consists of a single proton orbited by a single electron. A comparatively rare isotope, called deuterium, possesses the same chemical characteristics, yet has significantly greater mass because it contains one extra component: a neutron. Water made from 'heavy' hydrogen is barely distinguishable from normal water, except that one litre of light water weighs one kilo, while the heavy variety piles on an additional 100 grammes and has a boiling point of 101 degrees centigrade. 'Heavy water' is perfectly natural. Every swimming pool contains about one drinking tumbler's worth. But the heavy water is so thoroughly mixed in with the light that a sample of water extracted at random from the pool will yield fewer than a hundred heavy molecules out of every million. To be of use to researchers, the heavy water has to be extracted by electrolysis in a huge factory of tanks, distillation pipes and electrical generators.

By 1934 just such a plant was founded by the Norsk Hydro company at Rjukan in Telemark, southern Norway. When German troops occupied Norway in the spring of 1940, the plant was ordered to step up production. Once Einstein's warnings about the Nazi bomb programme had been taken seriously, the plant became a priority target for Allied raids. The problem was that the main building was protected by the steep flanks of a ravine, making it almost impossible to bomb from the air. Not that anyone on the Allied side relished carpet-bombing a Norwegian target. A more subtle sabotage mission seemed the better option.

In October 1942, a four-man team of Norwegians, trained in Britain by the secret Special Operations Executive, were parachuted back into their own country. It took some days for them to ski as unobtrusively as possible from their remote wilderness drop site to the Norsk plant, reconnoitre it and relay what they had discovered to the British. On 19

November, 30 men from the First Airborne Division were loaded into two giant gliders at a remote Scottish airfield and hauled aloft by a pair of Halifax bombers. The men had been trained over many weeks for sabotage attacks against an enemy factory, but were told only on the day of departure what their actual target was: the Norsk Hydro plant, atop its almost impossibly steep ravine flank. The plan was for the Halifax towplanes to release the gliders well away from the target, so that they could fly silently, and with any luck unobserved, through the Norwegian night, ready to drop their parachutists within striking distance. The empty gliders would then be allowed to crash. Unfortunately, navigation errors combined with atrocious weather to down the gliders prematurely, with their men still aboard. The Gestapo rounded up the survivors, all of whom were interrogated under torture, then shot.

At the Air Ministry in London, Reginald Jones was the young intelligence officer charged with deciding if another squad of saboteurs should be despatched. The decision, he later wrote, 'was even harder, knowing I would be safe in London, whatever happened to the second raid, and this seemed a singularly unfitting qualification for sending another 30 men to their deaths'. He reasoned that the Allied war effort had already decided that the Norsk Hydro plant had to be put out of action, or else why sacrifice the 30 men who had been lost already? *Not* to mount another attack would mean that their lives had been utterly wasted.

Norwegian resistance fighters had to be consulted first. Any attack on the plant would cause the occupying Germans to take reprisals among the civilian population of Vemorsk, the nearby village. Having been assured that the heavy water plant was genuinely valuable to the German war effort, the Norwegians agreed. This time a ten-strong squad of their own nationals would undertake the raid. They elected to make a long ski-trek to the target rather than risking an obtrusive air drop. Their journey took several weeks, but at last, on the night of 27 February 1943, they infiltrated the plant with the help of brave workers inside, then laid their explosive charges and escaped.

The small quantities of explosives in their backpacks could not inflict major structural damage on the heavy building itself. The charges were placed instead under the big water tanks, and at key pipework junctions. The resulting explosions caused the accumulated heavy water to spill into the factory floor drains and back, eventually, into the natural water cycle of the surrounding mountains and ravines. The saboteurs made sure to 'lose' a British sub-machine gun, dropping it on the factory floor to try to convince the Germans that this had been another British raid. The mission was a success, except that its small-scale explosions inflicted only minor damage. After three months the plant was back to normal. In November that same year, it was attacked by a 140-strong fleet of American B-17 bombers, which between them dropped 700 bombs. At least 600 missed the plant, tumbling uselessly into the deep ravine.

Even so, the aggressive scale of the raid persuaded the Germans that it was time to move shop. In February 1944 the heavy water equipment was dismantled and placed on board a civilian railway ferry ship, prior to being transported to Germany. Several members of the Norwegian sabotage team had been hiding in the snow-covered mountains throughout the past year, and with help from local partisans they blew up the ferry shortly after it set sail across Lake Tinnsjø. Oblivious to the lethal drama being staged around them, fourteen Norwegian passengers were killed. Everyone involved in this latest attack had known perfectly well that this might happen. The German Gestapo mounted the largest manhunt in its grim history for Knut Haukelid, the leader of both the sabotage raids. They never found him.

Unholy Trinity

By the mid-summer of 1945, the US War Department had prepared the text for a tremulously excitable press release, which in the event was not made public until after the war.

> Mounted on a steel tower, a revolutionary weapon
> destined to change war as we know it, or which may
> even be the instrumentality to end all wars, was set
> off with an impact which signalized man's entrance
> into a new physical world. Success was greater than
> the most ambitious estimates. A small amount of
> matter, the product of a chain of huge specially
> constructed industrial plants, was made to release
> the energy of the universe locked up within the atom
> from the beginning of time.

The world's first atomic explosion occurred on 16 July at a remote desert test site in New Mexico code named 'Trinity'. Leslie Grove's deputy, General Thomas Farrell, couldn't hide his visceral excitement: 'The effects could well be called unprecedented, magnificent, beautiful, stupendous and terrifying. No man-made phenomenon of such tremendous power had ever occurred before.' A more pithy comment came on the day from Kenneth Bainbridge, the scientist in charge of preparing the test site: 'Now we're all sons-of-bitches.'

Certainly this is the popular conception we have of Oppenheimer's reaction to his own achievements. In November 1947 he told an audience at the Massachusetts Institute of Technology that 'physicists have known sin, and this is a knowledge which they cannot lose'. Freeman Dyson, an English physicist who became closely involved with many of the bomb-makers after the war, thought carefully about these and other expressions of guilt from the Manhattan people. 'The sin of the physicists at Los Alamos did not lie in their having built a lethal weapon. They did not just build the bomb. They *enjoyed* building it. They had the best time of their lives building it. That, I believe, is what Oppenheimer had in mind when he said that they had sinned.'

Two decades after the event, Oppenheimer gave his impressions of the Trinity explosion during an emotional television interview for an NBC documentary, *The Decision to Drop the Bomb*. His continuing unease about what he had done seemed clear enough: 'We knew the world would not be the same. A few people laughed, a few people cried, most people were silent. I remembered the line from the Hindu scripture, the *Bhagavad-Gita*. Vishnu is trying to persuade the Prince that he should do his duty, and to impress him takes on his multi-armed form and says, "Now, I am become Death, the destroyer of worlds." I suppose we all felt that, one way or another.' Yet Oppenheimer's attitude to his creation could not be summed up as simplistic guilt. After he had presented a lecture in Geneva in 1964, one of his listeners asked: 'If you had foreseen the present situation in the world, would you have dared start the researches that led to the atomic bomb?'

'My role was to preside over an effort, to make, as soon as possible, something practical. But I would do it again.'

Another question from the floor: 'Given what has happened these past twenty years, would you – in the position you were in during the 1940s – would you again accept to develop the bomb?'

'To this I have answered yes.'

'Even after Hiroshima?'

'Yes.'

Teller's testimony

Oppenheimer's nemesis was a fellow Los Alamos scientist. Edward Teller was born into an affluent, educated Jewish family in Budapest, Hungary. As one of the great cities of the Austro-Hungarian Empire, Budapest was part of a larger central European world of predominantly German language and culture. In 1926, Teller left Budapest to study chemical engineering in Karlsruhe, Germany, as though he were simply segueing from one region of his natural home territory to the next. Two years later he transferred to the University of Munich to pursue his interest in chemistry. And then disaster struck. A streetcar accident cost him his right foot. Once he'd recovered from his injury and learned to walk with an artificial foot, he switched his allegiance to physics and transferred to the University of Leipzig, where he studied for a while under Werner Heisenberg. Teller might have settled down to a long, productive career in Germany, but like so many others of his background, he was forced to flee when Hitler came to power. In 1934 he joined Bohr's Institute in Copenhagen. A year later he went to America and stayed.

Teller's bitterness at his enforced exile bred in him a virulent hatred for fascism, and for its barely less palatable counterpart in the East, communism. In 1939, when Leo Szilard and Eugene Wigner had first warned Einstein about the possibilities of fission, Teller briefly became involved in the drafting of their letter to Roosevelt. Yet his attitude towards the bomb was vastly different to theirs, and he set about his work at Los Alamos with unveiled enthusiasm. If many of the physicists around him had any sense of reluctance about the possible outcome of the work they were doing, he did not share it. In fact he pestered Oppenheimer to start thinking about an even more powerful bomb: one that would emulate the heart of the sun, exploiting a process related to the 'fusion' of hydrogen atoms into helium with an accompanying massive release of energy. Teller's 'hydrogen bomb' was impossible for the Manhattan Project to pursue, and he made no effort to hide his frustration.

In 1949 America entered its most virulently self-destructive phase of anti-communist paranoia, spurred in large part by the rantings of an opportunist Republican senator, Joseph McCarthy. In the 1930s Depression years, many Americans had been sympathetic towards labour rights, stronger unions, and even communism. Oppenheimer was a leftist in the sense that he hoped for a better and fairer world than could be delivered by raw American capitalism. By the late 1940s most people who had fantasised that the Russian social experiment might be a way forward had been disillusioned. There could be no loving the mass murderer Stalin, although surely it must be possible to retain some elements of left-leaning sympathy without necessarily remaining a communist sympathiser? McCarthy and his unforgiving witch-hunters saw no such distinctions. Thousands of decent-minded liberal citizens were ensnared in a five-year nightmare of accusation, coercion and public disgrace, until the journalist Ed Murrow put a stop to McCarthy in 1954 simply by putting him on television so that all America could witness his unpleasantly hectoring style.

Oppenheimer's security clearance was revoked that same year. Teller had written to the Atomic Energy Commission (AEC) suggesting, disingenuously, that he wished Oppenheimer would be more supportive towards the hydrogen bomb. This was interpreted as an accusation of treachery, especially by the AEC's fervently anti-communist new director, Lewis Strauss, a man who disliked Oppenheimer personally as well as politically. Oppenheimer would not go quietly. Instead he demanded a formal hearing in an attempt to win back his clearance. Four weeks of interrogation began in April 1954. Forty witnesses were called, and most of them spoke favourably about him. Choosing his words with at least some measure of tact, the Manhattan Project's old military chief General Leslie Groves said that he might not have been inclined to grant Oppenheimer's security clearance if he had known earlier about his leftist sympathies, but he'd certainly had no complaints with his running of the Los Alamos team.

Teller's testimony was more overtly damaging. When the AEC prosecutor Roger Robb asked him if he wanted 'to suggest that Dr Oppenheimer is disloyal to the United States', he replied: 'I do not want to suggest anything of the kind.'

Robb rephrased his question: 'Do you or do you not believe that Dr Oppenheimer is a security risk?'

Teller then made the comment that destroyed any affection that might once have existed between him and the rest of the physics community, despite his subsequent fame as a key Cold War strategist and senior advisor to several presidents. 'Dr Oppenheimer acted in a way which for me was exceedingly hard to understand. I thoroughly disagreed with him in numerous issues and his actions frankly appeared to me confused and complicated. To this extent I feel that I would like to see the vital interests of this country in hands which I understand better, and therefore trust more.' With these words, Oppenheimer was effectively damned. The lone dissenting voice from within the AEC was that of Commissioner Henry DeWolf Smyth, formerly a physics professor at Princeton University, and now a respected advisor on atomic affairs. He concluded that 'there is no indication in the entire record that Dr Oppenheimer has ever divulged any secret information.' Despite more than a decade's worth of constant surveillance against him, 'supplemented by enthusiastic amateur help from powerful personal enemies', in Smyth's opinion Oppenheimer was not a potential traitor, just 'an able, imaginative human being with normal human weaknesses and failings'. Smyth's was a voice of reason in an unreasonable age – the age of the atomic bomb and its only possible hope: that Hiroshima and Nagasaki should be the beginning and the end of the bomb's use in war.

Although shaken, Oppenheimer was comforted by the respect of many colleagues. He found refuge at Princeton's Institute for Advanced Study, well away from the political maelstrom of nuclear policy. In 1963 a contrite AEC conferred on him its prestigious Enrico Fermi Award. Teller, meanwhile, learned to adjust his recollections as nebulously as Heisenberg remembering his humanitarianly motivated meetings with Bohr, or his supposed lack of zeal in trying to make a German bomb. Teller has sometimes been described as the real-life inspiration for the movie character Dr Strangelove. In the 1980s, as a key advisor to

President Ronald Reagan, he championed the development of anti-missile weapons for deployment in earth orbit: a highly controversial project that still has powerful allies today.

Pandora's box

The Manhattan Project stimulated a vast new nuclear industry which could not now be put back in its box. Wartime contingencies such as the Oak Ridge reactor and the Los Alamos laboratory became entrenched as permanent and fast-expanding institutions. University physics departments across America vied for funds to build ever bigger and more powerful cyclotrons and linear accelerators. Bench-top machines budgeted at a few hundred dollars in the 1930s now took up entire buildings and cost millions of dollars. Much of the money came from the US Department of Defense, keen now to stay ahead of Soviet Russian bomb-makers. Klaus Fuchs, a refugee German physicist, was employed by the British atomic energy programme during the Second World War and was sent to work on the Manhattan Project at Los Alamos, where he made significant technical contributions. None of his colleagues knew he was a committed communist who secretly passed detailed bomb documents to the Soviet Union. His activities weren't exposed until 1950. He was imprisoned in Britain for nine years, and then travelled to East Germany where he was appointed deputy director of a nuclear research institute. He was lucky not to have been caught by the Americans. Two New York-born communist sympathisers, Julius and Ethel Rosenberg, were convicted of passing atomic secrets to Russia, and both were executed by electric chair in 1953.

Niels Bohr also believed that Russia should have the bomb, although he never had the slightest intention of spying on Russia's behalf. Secrecy, he thought, was the problem, not the solution. As early as 1944 he had persuaded influential political figures in America and Britain that

Joseph Stalin should be told formally that the American bomb was nearly ready, and that the Soviet Union might be allowed to share in its control. Bohr hoped that this gesture of political openness would forestall a long, expensive and highly dangerous arms race. It was a brave stance. When he talked with President Roosevelt in 1944 he was met, initially at least, with sympathy and open-mindedness. But Britain's war leader Winston Churchill was overtly hostile. 'How did he come into the business? What's this all about? It seems to me Bohr ought to be confined, or at any rate made to see that he is very near the edge of mortal crimes.'

Just as Bohr had feared, Russian scientists lost little time replicating the American bomb. Their first explosive test was staged on 29 August 1949 at a remote site in Kazakhstan. The Cold War had begun, and both sides in the East–West divide were now solidly committed to an expensive and dangerous arms race; yet for all the secrecy and paranoia accompanying this drama, and notwithstanding the great armies of spies and counter-spies that each side unleashed upon the other, the subsequent 40-year nuclear stand-off actually *depended* on the knowledge that Bohr had wanted to transmit freely – the knowledge that both sides had the bomb and neither side could win. The Cold War was held in check for 40 years by the prospect of 'Mutually Assured Destruction', known throughout the nuclear weapons trade as MAD. This is what has saved us so far from the atom's darkest fires. Today the relatively simple strategies of the Cold War are long gone, and the balances of atomic power are not so certain. They could be perturbed at any moment.

PART SIX

Renormalising the Infinities

After 1945 the atom's energies could be unleashed for destruction, or harnessed and controlled as a source of energy. That choice was a matter for engineers and politicians. Meanwhile, theoretical physicists returned to their earlier conceptual problems. What was the atom?

By the spring of 1947, the once-great scientific institutions of Germany and Austria were in no fit state to conduct any research. Immediately after the war, a number of German physicists assembled in Berlin, hoping to reanimate the battered corpse of the Kaiser Wilhelm Institute for Physics. Max Planck agreed to serve once again as its president. Meanwhile, the British occupation authorities insisted on changing its name to the Max Planck Institute. Planck clearly was not associated with any of the Nazi atrocities. In 1933 he had pleaded with Hitler not to rid Germany of its fine Jewish scientists. The meeting was not a success, Hitler could not be reasoned with, and Planck recognised that his country was descending into hell.

One can only respect Planck's perseverance in trying to rescue some shards of honour from the wreckage of German science – and of his own life. His house was destroyed in February 1944 during a massive air raid. Every paper, every book and precious family memento was burned to ashes. In the last few months of the war, even as the Nazi regime began seriously to crumble, Planck's son Erwin was executed for his part in a failed plot to kill Hitler. Little wonder that the old man was exhausted. The resourceful Otto Hahn succeeded Planck as the Institute's chief in 1946. A few fragments of the once glittering Austro-German scientific jewel had survived, but for now it was hardly in the best shape to take the next steps in the atomic journey. There was still the Niels Bohr Institute in Copenhagen to serve as a focal point, an island of calm in a continent of massive displacements. Yet Bohr and Heisenberg were not much in contact after their eerie wartime meeting. Einstein had long departed from continental Europe, of course, as had Schrödinger, Fermi, Pauli and so many others.

In such epic times, when the fortunes of the entire world were being reshaped by the energies of the atom, there were thousands of jobs for ambitious physicists in America, and increasingly in Britain and France too. If there was a problem, it was that so many young physicists were rushing off straight after graduation to find work in the nuclear field: scaling up the laboratory accelerators, building bombs, designing civilian power stations or miniaturising the atomic piles to drive turbines inside ships and submarines.

The fashionable tool, now, was the particle accelerator, often known to the public as an 'atom smasher'. Imagine the Walton–Cockcroft machine on a ludicrously grand scale. Particles travel in a vacuum down a long copper tube, pushed along by waves made by powerful microwave generators surrounding the tube, called klystrons. Electromagnets running the length of the accelerator keep the particles confined in a narrow beam. When they strike a target at the end of the tunnel, various detectors record the events: the created particles and radiations released. Accelerators today are *huge*. The Stanford Linear Accelerator in California is 3 km long. Circular accelerators, such as the 27-km-diameter facility at CERN near Geneva, propel their particles around a circular track many times. As they sweep through any given section of the accelerator, the surrounding electric fields are modulated with extreme rapidity, so that the negative electric potential is always just behind where the electrons are, and the positive potential always just in front (or vice versa for positively charged protons). A particle accelerator isn't the delicate precision instrument that we might imagine. It's more like a blunderbuss than a rifle. Billions of particles are fired at billions of other particles, in the expectation that just a few will collide inside a detector, where their tracks can be recorded.

By the late 1940s this new and grandly-scaled technology threatened to swallow up thousands of physics graduates. There was a real danger that other types of physics – any concerned with things larger than atoms – might find themselves short of good people. So it might seem surprising when we encounter one physicist, Isaac Rabi, complaining in 1947 that 'the last eighteen years in our field have been the most sterile of the century'. In his opinion the last great highlight had obviously been the discovery of fission during the winter of 1938. He wasn't talking about the bombs and the power stations and accelerators. These were known devices, known pieces of hardware. They could be made more efficient now, more powerful and impressive. This was engineering, and it was all merely a question of money and application. People understood how the atom could be manipulated, steered, shattered and exploited. What they still didn't understand was how it *worked*.

If this seems strange, then think for a moment about the mobile phone. Who among us is not an expert at getting it to work? We text and email and voicemail and videophone as though it's second nature to us. We compare the costs of the service providers and worry about the memory chips and add-ons. We know the menus, and can even understand intuitively how the specs change from one manufacturer's phone to another, yet underneath they hold certain hierarchical similarities. All these things any ten-year-old kid *knows*. But how many of us really 'know' how a mobile phone works? And so it was with the atom after the bomb, the atomic power station and the nuclear-powered submarine. Everybody used it, but still no one really knew how it worked.

The science conference industry was thriving as engineers and technicians gathered, often many hundreds at a time, to talk about hardware and gossip about government research contracts. Amid this great buzz of optimism, a few mournful ghosts drifted lost and dismayed through the noisy hotel corridors and convention centre hallways. These were the deep thinkers, the ones who wanted to *know* more about the atom, rather than merely working out how to harness it for power. It was time for a new Solvay to create a forum where theoretical physicists could swap thoughts.

As so often in history, politics worked against a free international exchange of people and ideas, but in the spring of 1947, Duncan MacInnes, a former president of the New York Academy of Sciences, decided to give it a shot. His chosen venue was Shelter Island, New York, a couple of miles offshore from the forked eastern tip of Long Island. The antique Rams Head Inn hotel was chosen to set his delegates as free as possible from the usual noisy distractions of life. MacInnes asked only the topmost theoretical people to come, and his list amounted to fewer than 30 names. Oppenheimer was there, along with Hans Bethe, Edward Teller and several other of the most brilliant contributors to the Manhattan Project. Bohr and Einstein had been invited, but prior engagements kept them away. Other great figures from the pre-war quantum mechanics era were absent that day. Likewise, the necessary philosophical passions that they had brought to their science were somewhat depleted in the wake of the bomb's all too worldly bang. There was no Heisenberg, emphatic in his certainty about uncertainties; no tearful Schrödinger, no sarcastic Pauli, no taciturn Dirac ...

One of the often-observed truisms in physics is that a good theorist's work is essentially done by the time he or she is 30. Einstein was in his mid-twenties during his 'miracle year' of 1905, while Heisenberg had developed matrix mechanics by the time of his 25th birthday. By way of an exception, Schrödinger's wave equations emerged when he was 37 years old; but then, as one biographer has pointed out, he did enjoy the benefit of 'a late-flowering erotic out-burst' that seemed to rekindle his creativity. The truism remains true, however. Theoretical physics favours young minds above old ones. By the time of the Shelter Island conference, even Robert Oppenheimer, the 'father' of the latest and glaringly modern atomic phase of bombs and power, was in his mid-forties: young enough to do more work but not, perhaps, to create any more revolutions. It was time for a new generation of hotheads to take theoretical physics forwards. And there were some young talents at the conference, even if they were heavily outnumbered by their older sponsors. A young man named Richard Feynman was certainly keen to take up the reins and solve the latest impasse.

Surely you're *joking*, Mr Feynman!

Born in 1918, Feynman was raised in a quiet town called Far Rockaway, on the very outskirts of New York, near the sea. This was far enough away from the Big Apple that it was essentially a small American town in its own right; yet it was close enough that people there spoke with a distinct New York accent. Feynman would treasure that distinctive drawl throughout his life.

Richard's father Melville, a clever and articulate salesman and entrepreneur, tried his hand at various schemes, includ-ing a car polishing wax business, a chain of dry cleaning stores and a real estate company. The 1930s Depression hit him hard, but as America began gradually to militarise for a possible war, Melville finally lucked into a growing market: the uniform business. One of the first things he taught his son was that the uniforms that passed through his hands in their hundreds and thousands were just empty sacks of cloth, and especially the ones with fancy braid on them.

Richard learned from the earliest age never to trust outward appearances and always to question authority. He and his father pored over the *Encyclopaedia Britannica*, cutting through the verbiage written by some of the world's foremost authorities and 'translating back, back, back, into some way that I could understand'.

Melville had no particular scientific training in his own right, but his curiosity about the world was striking. He had a knack for conveying facts in terms of images and ideas that might make sense to a young boy. When Richard read, for instance, that a certain dinosaur had been 25 feet high, with a head six feet across, Melville explained that such a creature would have been tall enough to stare through Richard's top floor bedroom window, but wouldn't have been able to get its head into the room without smashing the window frame. Richard could now understand the impressive scale of the dinosaur in a way that the more exact numerical description had failed to convey. One day he would tackle some of the most complicated ideas in physics in a similarly robust and accessible style.

The teenaged Feynman was fascinated by radios, which in the 1930s were built from distinct circuits, capacitors, amplifiers, tuners and valves that you could actually get your hands on and fiddle with. Their internal logic was not hidden as it is now by blank-faced microchips in sealed plastic pods. Feynman liked to impress his neighbours by listening to the strange wheeps and squeals of a faulty radio, analysing the problem in his head, and only then reaching for the screwdriver, taking off the back cover and unhesitatingly swapping around a couple of wrongly installed valves. He gained a reputation as 'the kid who fixes radios just by thinking!'

At high school, Feynman's approach to maths and physics – he really wasn't interested in much else except girls – was to read the first few paragraphs of a text book explanation, then throw down the book and start writing out some ideas of his own. Some of his teachers thought it strange that he should waste so much time trying to recalculate what the great physicists of the past had already figured out, but Feynman refused be told how things worked, or what was true or false, or why a certain equation mattered. He needed to discover a thing for himself, because only then would he be able to *understand* it. He spent the first few years of his formal education working out principles that everyone around him already thought they knew. He didn't care. Every time he found something out for himself, it was 'the kick in the discovery' that thrilled him. As he headed to MIT and the beginnings of his career in theoretical physics, he kept up this rather flippant pattern, not even properly reading the freshest papers with all the newest theories and experimental results. He'd just skim the opening arguments, then toss them aside. The only papers he did read thoroughly, and show any respectful appreciation towards, were Paul Dirac's masterful synthesis of Heisenberg's matrix mechanics and Schrödinger's wave functions.

When it came time for graduate school, Feynman thought he'd just stay on at MIT, but his department head, John Slater, wanted him to widen his horizons. He urged that Feynman should go to Princeton, and wrote a glowing letter of commendation to that effect. But the career progression for young men such as he was not so simple in those days. Princeton's head of physics, Henry DeWolf Smyth, wrote back saying that 'one question always arises, particularly with men interested in theoretical physics. Is Feynman Jewish? We have no definite rules against Jews but have to keep their proportion in our department reasonably small because of the difficulty of placing them.' Smyth's concern, apparently, was that Jews would find it unusually hard to find decent jobs once they'd left Princeton. Not that the University had any objections to Jewish people itself, of course. It was a subtle hypocrisy. Slater's recommendations swayed the case, and Feynman arrived at Princeton in the autumn of 1939. His first duty was to attend a stuffy tea party hosted by the dean of the graduate school – or rather, by his socially alert wife. Almost as soon as Feynman had walked into the room, she asked him: 'Would you like cream or lemon in your tea, Mr Feynman?'

Not having encountered such a strange choice before (he was a coffee-drinker anyway), he absent-mindedly replied: 'I'll have both, thank you.'

'Surely you're *joking*, Mr Feynman!' his hostess teased, making it quite clear that he had committed a lower-class faux pas in front of all her genteel, well-mannered Princeton colleagues.

Feynman didn't give a damn. He flaunted his blue-collar, small-town mannerisms as a badge of disdain for pomp, hypocrisy, class and racial prejudices and other such 'hocus-pocus' that had no place in a clear-thinking mind. He sneered at the airy pretensions of middle-class art, literature, poetry and music, not to mention the world's great canon of 'philawzaphers', all of whom, he thought, were just messing about with words. He wasn't much impressed, for instance, by Descartes' famous phrase, 'I think, therefore I am'. When *he* thought about it, he couldn't read any more into it than 'I am, and I also think'. If that was philosophy, then it wasn't much use as a tool for finding out about the world, he reckoned. Of course, much of this flippancy was just an act, a flamboyant gesture from a man much given to bravura storytelling. As he matured, Feynman sometimes forgot his larger-than-life persona and often spoke as wisely as any great 'philawzapher' about the meanings of science and its relationship to the broader culture. And he learned to play the bongos, also forgetting his distaste for music.

While at Princeton, Feynman developed what would be a lifelong fascination for a concept in physics known as 'the principle of least action'. It has a touch of magic about it:

A ball at the top of a tall building has potential energy. When the ball is tossed out of the window and starts plunging down to earth, its potential energy is gradually used up while the kinetic energy is growing. By the time the ball is just about to hit the ground, the potential energy – the 'work' that the ball *could* do – has almost entirely been converted into kinetic energy, the work that the ball actually *is* doing. Feynman was intrigued to discover that the flying ball is always 'looking for' the most efficient balance that it can find between lost potential energy and gained kinetic energy. To put this rather mathematical idea even more bluntly, the ball finds the least effortful path to the ground. The same applies to orbiting satellites and moons, or apples falling from trees, or laser beams wending their way in and out of complex prisms and lenses. Inanimate things, from cannonballs to photons of light, apparently 'know' how to take the most efficient paths from their start position to their end one.

Our ordinary understanding of this is muddled by our insistent belief in the 'path' that seems to carry those things from point A to point B. Feynman knew from high school the neat traditional diagrams of balls making graceful ballistic arcs across the sky. Now he thought instead about what was going on at any given point of that arc. He saw that the arc itself was an abstraction, a meaningless pencil sweep across a page. All that counted was the ball's microsecond-by-microsecond 'decision' about what to do next. He wondered if some of the abstractions of quantum physics could be resolved in a similar way. In his emerging concept, there were no electron waves to worry about. They were just abstractions, no more real than the sweeping arc of a ball's flight. Similarly he didn't worry about the possible shapes or paths that might apply to an electron's journey. It could zig-zag or turn loop-the-loops for all he cared. Again, just abstractions. You just looked at the two things you could ever know for sure about an electron *after* doing an experiment: its starting point and its end point. Then you worked out all the myriad paths that the electron could have taken, and isolated the far fewer paths that would have burdened the electron with the least amount of work to do as it navigated through your experiment. It was a reformulation of the weirder probability aspects of quantum mechanics in a language reasonably familiar to classical physicists. The English mathematician William Hamilton had explored the principle of least action more than a century earlier. Feynman's new 'kick' was to apply those ideas to the shimmering uncertainties of the atom, the electron and the photon.

Feynman's mentor at Princeton, the renowned physicist John Wheeler, was impressed by his precocious young student: 'He treats on a footing of absolute equality every conceivable history that leads from the initial state to the final one, no matter how crazy the motion in between. The contributions of these histories are nothing but the classic [principle of least] action integral. How could one ever want a simpler way to see what quantum theory is all about!'

Even as he began to make his mark as an original theorist, Feynman's personal life was in turmoil. Arline Greenbaum, his much-beloved girlfriend, had been diagnosed with tuberculosis. Today this disease can be effectively treated with antibiotics. In the early 1940s it was still essentially a lingering death sentence. D.H. Lawrence had died from it, along with countless other doomed heroes and heroines in literature, both fictive and real. Operatic courtesans expired sweetly from it on stage, but the reality was not so romantic for Feynman or his lover. Tuberculosis carried a stigma, rather as the HIV infection does today. In June 1942, Richard and Arline eloped and got married, despite opposition from all quarters. Lucille Feynman wrote a harsh letter to her son: 'Your marriage at this time seems a selfish thing to do, just to please one person. I was surprised to learn such a marriage is not unlawful. It ought to be.' It would be a long while before Richard could forgive her for that.

Meanwhile, America's secret bomb project was gaining momentum. Robert Oppenheimer hoped to recruit Feynman but knew he was unlikely to want to work for a military project – too many uniforms, too many people telling him what to do – so he played the one card guaranteed to win Feynman's trust. He said there was a sanatorium near Los Alamos where Arline would be as comfortable as it was possible for her to be. It would be easy for Feynman to visit her every weekend. And so he went to Los Alamos and played his part in the making of the bomb. He impressed Oppenheimer as 'the most brilliant young physicist here. He is a man with a thoroughly engaging character and personality.' It would be hard to say that any one part of the bomb relied crucially on Feynman's work. It was more that he acted as a conduit for ideas, a sounding board for problems. He was a catalyst around which some truly great work was done in those years. Yet he knew that his particular mark on physics had yet to be made.

'They don't know'

Arline died on 16 June 1945. Feynman lived his life in an unhappy daze for the next four weeks. He was momentarily thrilled by the first test of the atom bomb at the Trinity site on 16 July, when he insisted on watching the blast with his naked eyes instead of wearing the prescribed darkened goggles. He felt there was nothing more for him to do at Los Alamos. A job offer arrived from Cornell University at Ithaca, New York. Feynman accepted readily enough, and worked alongside his Los Alamos colleague Hans Bethe. A period of ferocious womanising distracted him from his grief, and from the sense that he was pretty much finished as a physicist – that he hadn't anything truly original to contribute from here on. His morale was at its lowest ebb, although his good friend Bethe couldn't help observing that 'Feynman depressed is just a little more cheerful than any other person when he is exuberant.'

The bomb preoccupied him, coloured his every waking moment – at least, those moments that weren't taken up trying and failing to compensate for the absence of Arline. He made a cautious peace with his mother, especially when she admitted how much she'd admired Arline's courage. But as they sipped coffee in a New York diner one day, all Feynman could think about was the blast radius from a nuclear attack, and how much of the city around them would be destroyed. He saw construction workers putting up new skyscrapers, or repairing the city's great bridges, and he said to himself: 'They don't know. They don't know, and they're all wasting their time. There's no point.' Feynman's honesty was to recognise that the bomb threatened the safety of all mankind, and he'd played a part in making it happen. He was also honest about why he'd participated. He was having too much fun. He was just too damned intellectually *curious* to see if this thing could be done. The bombing of Japan in 1945 was not what he and the others at Los Alamos had expected when they first set to work in 1942, supposing themselves to be in competition with Heisenberg and the German Nazi regime. 'What I did – immorally, I'd say – was not to remember the reason that I said I was doing it. So when the reason changed after Germany was defeated, not a single thought came into my head about what that meant. I simply didn't think.' His error, now, was to imagine that he carried the burden of the atomic age on his shoulders alone.

And then one day he walked into the cafeteria back at Cornell and saw 'some guy fooling around, throwing a plate into the air'. His mind suddenly sparked into life again. There was the obvious circular rotation of the plate, which anyone could see, and a less obvious wobble that he had to analyse a little more carefully. There were two distinct kinds of momentum going on here, a rhythmic and cyclical up and down one as well as the simpler rotational one. On a whim he decided to do the maths and see what came out. It was 'just for the fun of it', he thought. Actually, when he looked at the thing a little more over the next few days he thought he could sense some clues about the complicated spin of electrons ... There was nothing especially original about his musings on the plate, but he was back in business, and his curiosity about the atom had reawakened with a vengeance.

Different infinities

One of the great challenges facing post-war physicists was the problem of the relationship between energy and matter: photons of electromagnetism coming into contact with – or being generated by – the particles of the atom. The emerging science of these interactions was called Quantum Electrodynamics, or QED. It was dominated by complicated descriptions of 'fields'. We are so used to talking about magnetic fields, gravitational fields and so forth that it's worth stopping for a moment to look at this somewhat slippery concept.

Isaac Newton described a universe where gravity was an invisible force acting between objects, causing masses (apples, planets, stars, galaxies) to pull towards each other even when separated by a void. He was not able to explain what gravity actually was, yet he was well aware that an explanation was needed. Today that mystery remains more or less intact, although Einstein has given us an alternative way of talking about gravity. It's a curvature of space, a 'field' caused by the distorting presence of mass in space-time. Electromagnetic fields have something of the flavour of gravity about them, except that they can repel other electromagnetic objects as well as drawing them in.

Are fields real? In 1850, Michael Faraday made 'iron filing diagrammes' to illustrate that the empty space immediately surrounding a magnet had a certain structure. He scattered iron filings onto sheets of paper impregnated with wax. Tapping the paper caused the filings to arrange themselves neatly along lines of force surrounding the magnet like a ghostly aura. Then a gentle heating of the wax glued the filings into place. The first pictures of invisible fields had been created. The results, preserved to this day at the Royal Institution in London, seemed to prove the existence of magnetic fields, but the picture is not so straightforward. What we actually see are the products of the iron filings' interactions with a nearby magnet. There's no proof that magnetic fields (or gravitational fields, for that matter) exist as entities in their own right, independently of the things that are affected by them, thus causing some kind of measurable event to take place. Faraday's 150-year-old wax sheets present us still with a profound mystery. Not only can't we be sure of what we are seeing, we can't even be sure if it's really there in the first place.

The post-war atom was replete with fields. The charges of the electron and proton operated as fields, Dirac's electrons and positrons popped in and out of existence from fields, and so on. Everyone was used to talking about electro-magnetic fields surrounding wires and magnets and radio waves. All the nebulous and wavy paraphernalia of late 19th-century science had worked with breathtaking success, but could the same language apply to atoms?

The fudge between particle and wave descriptions for the electron remained more than just a philosophical inheritance from pre-war quantum mechanics. If an electron has mass (albeit very small) and electric charge, then these qualities, *exerted as fields*, act not just on other subatomic particles but on the electron itself, just as earth's gravity acts on the crust of the planet, as well as on orbiting spacecraft or other celestial bodies like the moon. And just as the diameter of the earth counts in any description of its gravity field, so it's equally important to know what 'diameter' an electron has when it's being treated as a particle in the calculations, or when it's detected, bullet-like, in a bubble chamber or some other experiment. If it's extremely small, then the forces exerted by the electron *on itself* can be calculated and factored into all the quantum mechanics equations. If an electron in its particle guise is just a point with zero diameter, then all the equations go crazy.

Think of a hammer banging down on a block of steel. It's easy to work out from the momentum of the hammer, and the surface area of the steel target, how much energy the block absorbs when the hammer hits it. Simply divide the energy of the hammer blow by the topmost surface area of the steel and we have our answer. Now think of the hammer smashing down with exactly the same force onto the tip of a strong but very slender metal nail instead of a block. The hammer blow is applied to a much smaller area, but it's still possible to calculate exactly how much force gets exerted on the nail. What happens, however, if the nail is so slender it has zero diameter? Then the mathematics gives a strange result. The hammer apparently comes down with infinite force on a nailhead with an infinitely small surface area.

Electrons exert their negative charge on other particles, especially by repelling other electrons. That same charge also acts on the electrons themselves. If electrons in their particle guise are thought of as point-like entities with zero diameter, then the 'self-energy' of the charge on its own electron becomes infinite. To put it another way, an electron's negative charge should repel itself with infinite force! The same problem applies to an electron's mass. Concentrated in an infinitely small volume, it threatens to become infinitely dense, and so on and so on. And so the physicists at the 1947 Shelter Island conference gloomily contemplated the plethora of inconvenient infinities that were cropping up on either side of their equations from the 'self-energy' of infinitely tiny electrons. Standard procedure at that time was simply to cancel them out in a mathematical procedure called 'renormalisation'.

Suppose we are calculating an equation with an inconveniently huge number on both sides of the '=' sign:

$$10 \times 25 \times 1{,}000{,}000{,}000 = 5 \times 50 \times 1{,}000{,}000{,}000$$

It's tempting just to eliminate the 1,000,000,000 from both sides of the equation to reveal a neater and simpler equation which still gives the same result in the end:

$$10 \times 25 = 5 \times 50$$

Suppose that 'inconveniently huge number' is infinity, and the same manoeuvre can be applied to get rid of it. The snag, though, is that infinity does not always equal infinity. Think of all the whole numbers that exist: 1, 2, 3, 4 and so on, in a never-ending series stretching to infinity. Now think just of the even numbers: 2, 4, 6, 8 … Only half of all the whole numbers in the infinite series of numbers can be even. So surely that means that there are *two* kinds of infinity here, with one of them being twice as big as the other? The atomic theorists crossed their fingers and hoped that the infinities in their calculations were all the same kind of infinity. Everyone knew that renormalisation was nothing but a mathematical conjuring trick, a piece of 'hocus-pocus', as Feynman called it.

The mathematically austere Paul Dirac was offended by the whole thing, and had spent his last years, somewhat like Einstein, in a fruitless search for a more satisfying alternative to something he didn't like, even though the rest of the physics community accepted renormalisation because it appeared to work. 'Sensible mathematics involves neglecting a quantity when it turns out to be small – *not* neglecting it just because it is infinitely great and you do not want it!' he said. Another option, assigning an extremely small but specific value to the diameter of an electron, at least gave the equations a finite number to manipulate, yet this value was just an invention, a make-believe repair patch. Its name, the 'classical electron radius', just shows what a desperate ploy it was.

At the Shelter Island conference, brash 28-year-old Richard Feynman thought he could see the beginnings of an answer. What if the interactions between particles were not governed by fields or forces? What if particles communicated *all* their influences – mass, charge, energy – by means of other particles? Feynman thought that the problem lay with physicists' talk about fields. Everyone was so worked up about the *area* over which a field worked, and how small or how large the electron was. What, Feynman wondered, if there *was* no field? What if there were only particles? Somewhat like Heisenberg, he chose to concentrate only on the starting points and end points of a particle's history, as we've seen. Whatever happened in between, whether or not it was mediated by some kind of a field, was after all just another abstraction that could never be seen in any experiment. He thought about each subatomic particle *only* in terms of its effect on another. He also decided that individual particles on their own were essentially meaningless. If a particle did something, then there would have to be another particle for it to *do something to*. Feynman's next piece of bravura thinking was that once you'd found a link of action between one particle and another, it didn't matter which way you ran the film, so to speak. His interactions worked backwards in time as well as forwards. From this logic, he determined that Dirac's anti-matter positrons were electrons travelling backwards in time …

Feynman explained this idea with a beguiling visual image. Suppose a thin piece of string, winding this way and that in loops and spirals, is embedded in a block of plastic. The string represents a particle's possible motions while it's in a state of (unobserved) quantum uncertainty. There's nothing predictable about which way the string will go, or where it will wend and wind. The block is opaque, so you can't see the string anyway. Now suppose an experimenter makes an observation at a particular moment in time, as represented by slicing the plastic block at a certain point. The cross-section may reveal one tiny dot, where the string inside has been neatly severed. That's a particle. But if the slice-through reveals two little black dots of string, one apparently with a 'clockwise' bias to the twists of its individual fibres, and another with an anti-clockwise twist, you might be inclined to think that you were dealing with two 'opposite' particles. According to Feynman, they are the same particle caught going forwards *and* backwards in time.

Feynman wanted to cut through the infinities, and eliminate the complexities of talking about fields, by reducing electron behaviour to just three basic ideas. 1) An electron moves through space. 2) A photon moves through space. 3) An electron either emits or absorbs a photon. And that was pretty much it. You didn't worry about the interactions or distributions of abstract fields. They were as insubstantial and meaningless as those 'arcs' described by falling cannonballs. You concentrated only on the exchange of photons between particles. And you didn't worry about going backwards in time. Finally, if you wanted to work out the charge of an electron, or its mass, then you just took the best available data from the accelerator experiments and fed them in. They were already good enough to one part in 100 million, so why waste any more time trying to calculate the exact figures? Mathematical pedantry just generated all those infinities. Feynman didn't care about the niceties, which could be worked out later. He compared the problem to trying to work out the exact diagonal of a 5-foot square. Given that the answer yields the square root of 50, it's an 'irrational' number with an infinite series of decimal places: 7.071067811... 'It's not philosophy we are after, but the behaviour of real things. So in despair, I measure the diagonal directly – and look, it's *near* to seven feet – neither infinity nor zero. So we've measured these things [electron masses and charges] for which our theory gives such absurd answers.'

With this ruthlessly simple reasoning, Feynman came up with a tool kit of interactions that worked, albeit at an aggressively practical level that left the seekers after ultimate truth somewhat dissatisfied. If an electron did have a specific mass, they wanted to know it exactly. It also had to follow from the theories *why* it had that mass and no other. Feynman was unconcerned for now, so long as his technique delivered useable results with an accuracy that could be tested in the laboratory. He substituted swathes of complex mathematical reasoning for 'Feynman diagrams' that summed up just the particle interactions and what they could produce.

But his first halting attempts to get some of these ideas

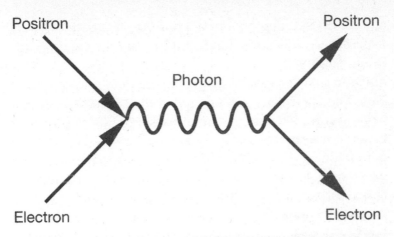

Figure 4. A typical Feynman diagram showing an electron and its anti-matter counterpart, a positron, colliding to create a photon of gamma radiation (wavy line).

across to the Shelter Island conference were not entirely successful. A second gathering in 1948 at Pocono in Pennsylvania went even more badly for him. Freeman Dyson, an English physicist who was soon to play a significant role in Feynman's life, sympathised with the brash New Yorker's difficulties: 'The reasons his ideas were so hard for people to grasp was that he didn't use equations. He had a physical picture of the way things happen ... Other people's minds were analytical. His was pictorial.' Feynman was having a hard time making his audience see that his diagrams were just visual tools, just another and slightly simpler way of describing an electron's behaviour. They weren't actually supposed to be pictures of *real* things. When he drew a neat straight line symbolising the path of an electron, for instance, just before it got slammed by a photon coming in from the side, Niels Bohr complained that he obviously hadn't understood the first thing about the uncertainty of an electron's path! The grand old men of quantum mechanics had spent their finest years arguing about how to picture electrons, or else claiming that the only sensible description was a purely mathematical one. Feynman had devised a neat visual shorthand for talking about electron–photon interactions. The fact that it looked, superficially, like some kind of a 'picture' threw everyone into confusion.

If you look at a chart showing the rise of the population of New York between 1900 and 2006, that's a visual aid that helps you understand and manipulate the data. It's much easier to interpret than a long and hard-to-memorise list of population figures and calendar years. Statisticians regularly make such pictures, yet no one would ever confuse a pie chart or a bar chart with the actual population of New York. In similar vein, Feynman's clever diagrams clarified the complicated interactions between electrons and photons. But they *weren't* pictures. It took a while for some of the Old Guard of quantum mechanics to understand that distinction. Feynman was patient with them, because he too was worried that his solutions might be just 'a half-assedly thought-out pictorial semi-vision thing. I would see the jiggle-jiggle-jiggle or wiggle of the path. It's all visual. It's hard to explain.'

A clash of styles

Feynman's main competitor at that time, Julian Schwinger, was also working on the infinities problem, using complicated mathematical logic to try to eliminate them without any fast and loose conjuring tricks. Feynman's presentations were colourful, energetic, spontaneous. His arms, his legs, his whole body seemed to enact the drama, and his broad, uncompromising New York accent gave him the aura of a frenetic salesman who really believed in his product. Schwinger, a product of an upmarket childhood among Manhattan's fashion socialites, was more precise in his language, and his presentations were made almost entirely in algebraic formulae. It was a clash of styles: Feynman's loose blue-collar analogies alienated his audience of formal thinkers, while Schwinger couldn't climb down from his privileged cleverness, and lost anyone whose mathematics wasn't as good as his, even if they remained impressed by his dazzling skill. As the delegates left the 1947 and 1948 Shelter Island and Pocono conferences, few of them had quite grasped how Schwinger's complicated equations worked, and even fewer thought that Feynman's crazy ideas would hold water. Even so, there was a palpable feeling that the theoretical physics of the atom was about to climb out of the pit where it had been bogged down for the last two decades.

The emerging drama had some of the flavour of the Heisenberg–Schrödinger debates back in 1925. On one side there was a reckless womaniser and bon viveur peddling a simple idea of electron–photon exchanges that could be used constructively by anyone versed in quantum mechanics; on the other, there was a proud logician looking only at what was mathematically justifiable, and if his reasoning was too algebraically sophisticated for people to follow, then too bad. And just as Schrödinger's and Heisenberg's ideas had been merged by the English physicist Paul Dirac, so Feynman's and Schwinger's totally different approaches were reconciled, now, by another Englishman, Freeman Dyson. But where Dirac had been socially gauche, and often curt to the point of rudeness in his conversations, Dyson was the politest young fellow you could hope to meet. He and Feynman became good friends.

Tempted by Mephistopheles

Dyson was what's known today as a 'hothouse' child, reared in an 'accelerated learning environment' by parents keen for him to succeed. At eight years of age he was sent to board at Twyford College, which he found a 'strange and forbidding' place, albeit geared towards academic excellence. The school was just three miles from home, yet Dyson's parents seldom visited him in term time. They meant no reproach by this. His father, George, was a composer and music teacher who eventually became director of the prestigious Royal College of Music. His mother, Mildred, had studied and practised law, and now passed on to her son her deep love of literature. This was no loveless Dirac household, but even so, he did grow up with some of the shyness and reserve we might today associate with a 'brainiac' or a 'nerd' – terms we simplistically use for people who are clever in matters of the mind yet who cannot deal with some of the broader aspects of life.

Then came Winchester, one of Britain's most successful public schools. A fellow pupil remembers Dyson as 'a tall slightly-built boy with piercing eyes and an infectious, slightly sardonic laugh'. Here he began reading Dirac's quantum equations while his classmates were still working out how to do calculus. As he remembered many years later: 'There was lots of stuff [in the library] about electrons and electricity and radio waves and all sorts of things. I remember asking people, "Why is it that they only talk about electrons and not about protons?" Nobody seemed to know.' In keeping with so many of the great physicists, Dyson began asking awkward questions from the earliest age, which few of his seniors could answer. Even his parents were becoming a little anxious that their bright boy might be flying too close to the sun. His mother took him for a walk one day during the Christmas holidays, and she told him the story of Goethe's Faust, a man who sells his soul to the Devil in return for knowledge.

In 1941 Dyson went up to Trinity College, Cambridge on a scholarship, where his self-confidence blossomed somewhat (his favourite sport was to climb the buildings late at night). His darkest period came in the mid-1940s, working as a researcher for the Royal Air Force's Bomber Command. The illogical mess of the human world could not have been more different from the clarity of pure reason, he discovered. One of his first tasks was to analyse the safety figures for air crews bailing out of stricken bombers. He totted up the aircraft losses, then compared those numbers with the crew survival figures. A great many crews were lost in their entirety, even though bombers did not necessarily blow up the instant they were attacked. Surely some of the men should be able to bail out? Dyson inspected planes on the ground, and decided that the first man to try to bail out of a stricken plane tended to get stuck in the narrow escape hatch, dooming himself and his companions.

Dyson also found that defensive gun turrets made bombers heavier and slower, and needed more crewmen to operate them (putting more men at risk per mission). The net result was that gun turrets made no difference to a crewman's overall chances of survival. Finally he looked at the massive bombing raids against enemy targets, and concluded that they were ineffective, unlikely to hasten the war's end, and not worth sacrificing the lives of so many airmen. By now Bomber Command was unwilling to address any of these problems in a spirit of honesty. It could not admit to its aircrews, for instance, that such tragic mistakes had been made. The firebombing of Dresden and Hamburg plunged Dyson into the Faustian hell his mother had warned him about. He decided that 'Bomber Command might have been invented by some sociologist to exhibit as clearly as possible the evil aspects of science and technology.'

When the war came to an end, Dyson gladly accepted the chance to go to Cornell, where Hans Bethe was keen to take him on as a graduate student. Here he encountered Feynman and began his education in the more cheerful possibilities of human experience. In particular there was a road trip to Albuquerque, where Feynman hoped to chase down an old girlfriend. There were incidents with cops, a tenacious pair of hitchhikers, an emergency sleepover in a flophouse brothel (Feynman said all the respectable hotels were full) and other chaotic pieces of nonsense that certainly helped Dyson to live by one of his mother's favourite sayings, taken from a 2,000-year-old poem in Latin by Publius Terentius: 'Homo sum; humani a me nihil alienum puto.' I am human; nothing human is alien to me.

After this eye-opening adventure, Feynman went off to

find his girlfriend, and Dyson decided to explore more of America by Greyhound bus. Three days later, while he was trundling uneventfully across the Nebraska plains, 'something suddenly happened. For two weeks I had not thought about physics, and now it came bursting into my consciousness like an explosion. Feynman's pictures and Schwinger's equations began sorting themselves out in my head. They were just looking at the same set of ideas from two different sides.' In the summer of 1948 he prepared the first of his 'theories of the middle ground', slightly nervous because neither Feynman nor Schwinger had officially published *their* ideas yet. He had discussed his qualms with Bethe, who advised that Feynman would probably be alright about it, because he habitually couldn't be bothered sending papers off anyway. But Schwinger might be touchy. Dyson decided it was their fault for not publishing, and sent his paper to the *Physical Review* for publication. Feynman was delighted for him – although he liked to pretend that he couldn't be bothered to read the paper because, of course, *he* didn't need to. Schwinger wasn't quite so pleased to see his clever maths outshone by a subordinate youngster and his flashy collaborator, although typically his response was too abstract for anyone but him to understand: 'There are visions at large, being proclaimed in a manner somewhat akin to that of the Apostles, who used Greek logic to bring the Hebrew god to the Gentiles.'

No contest

As soon as the 1948 Pocono conference ended, Oppenheimer returned to Princeton moderately intrigued by Schwinger's calculations, and entirely unmoved by Feynman's arm-waving and diagrammatic flourishes. Waiting for him on his desk was a letter from a Japanese theorist that reawakened his interest. 'I have taken the liberty of sending you several copies of papers and notes ...' Shin-Itero Tomonaga had thought about similar problems just at a time when his country's communications with the outside world were almost completely cut off. He had actually *anticipated* Schwinger's ideas about renormalisation by several years. Obviously there was something here after all – and Tomonaga's maths was a good deal easier to comprehend than Schwinger's. Now that Japan's relationship with America was returning to something like normality, Oppenheimer lost no time inviting Tomonaga to Princeton and back into the fold. On meeting Tomonaga, Dyson was struck by his uncomplaining dignity: 'He is more able than either Schwinger or Feynman to talk about ideas other than his own. And he has enough of his own, too. He is an exceptionally unselfish person.'

In the autumn of 1948, Dyson, a junior young voice in a field of giants, launched a brazenly fearless campaign on Feynman's behalf. He wrote to Oppenheimer practically demanding that he take an interest: 'I am convinced that the Feynman theory is considerably easier to use, understand, and teach. We have a new theory of nuclear fields which can be developed to the point where it can be compared with experiment, and this is a challenge to be accepted with enthusiasm.' Oppenheimer set up a series of meetings so that Dyson could make his case. Caught somewhat by surprise, Feynman was stuck at Cornell, and did not attend the Princeton showdowns. He had to follow Dyson's progress via long-distance phone calls. Hans Bethe was a kindly presence, travelling down from New York specially to lend moral support on Feynman's behalf. Oppenheimer was a furious critic, chain smoking, jabbing his finger in the air at every slight error in Dyson's presentation, and generally behaving like a nervous and ill-disposed customer. Bethe, meanwhile, let everyone know that he was on Dyson and Feynman's side. He then took Oppenheimer for a quiet chat in private, and suddenly the battle was won. QED was in the hands of Feynman and Dyson now.

A few months later, both men attended a conference at the American Physical Society, where many of the delegates spoke of the 'Feynman–Dyson' theory or referred to flipcharts of 'Feynman diagrams' as if these ideas had been solidly entrenched for years. Feynman turned to his young champion and said: 'Well, Doc, you're in!' Dyson hadn't actually earned a Doctorate, but he'd certainly won his spurs. The prize he most valued was a one-line handwritten note from Oppenheimer after the dust from the dispute had settled. '*Nolo Contendere*. R.O.' No contest.

The QED theory of interactions between electrons and photons had reached an incredible degree of accuracy. It was as if a team of navigators had worked out the distance between Los Angeles and New York to within a single hair's breadth. QED became the standard parlance, and in 1965 Schwinger, Feynman and Tomonaga were awarded the Nobel Prize 'for their fundamental work in quantum electrodynamics, with deep-ploughing consequences for the physics of elementary particles'. Schwinger was pleased, of course, yet he never lost his distaste for Feynman's shortcuts. Two decades later, he compared Feynman diagrams to the hand-held electronic calculators which students were allowed, by then, to take into exam rooms: 'Like the silicon chip of more recent years, the Feynman diagram was bringing computation to the masses.' He did not mean this as a compliment.

Dyson's invaluable contributions were not honoured in Stockholm that year, possibly because he had been only a young student at the time of his most significant work, but more likely because of the Nobel Committee's awkward regulation that no more than three people can share an award. Dyson was not being bitter when he wrote: 'Tomonaga, Schwinger, and Feynman rescued the QED theory without making any radical innovations. Theirs was a victory of conservation. They kept the physical basis of the theory … precisely as it had been laid down by Dirac, and only changed the mathematical superstructure.' Serious problems had been ironed out, yet as we'll soon see, the atom contained many more mysteries.

The atomic spaceship

Dyson's next adventure was the perfect expression of something we have almost forgotten today. Even in the shadow of the bomb, the world believed for a moment that 'the friendly atom' might yet revolutionise our lives for the better. There would be atomic cars and trains and planes, and of course, atomic spaceships. Project Orion was an engineering study for a spacecraft powered by nuclear pulse propulsion, an idea first proposed by Stanislaw Ulam in 1947. Initiated in 1958, Orion was led by Theodore Taylor at the General Atomics company, with Dyson as his principal consultant. Both men were convinced that chemical rockets, with their limited payloads and high cost, were the wrong approach to space travel. Orion, they argued, was simple, and above all, affordable. Taylor even proposed that the vehicle be launched from the ground. The ship was expected to be sixteen storeys high, 10,000 tons in weight, shaped like the tip of a bullet. The engine consisted of a flat, circular 'pusher plate' at the base, with a small hole in its centre, through which miniaturised nuclear bomblets, 'pulse units', were to be ejected at the rate of one every second. Huge shock absorbers between the pusher plate and the main body of the ship evened out the staccato stresses of the bomb blasts, and the great monster was supposed to rise into orbit at the rate of one explosion per second. After that, it would head for the moon or Mars at a more modest rate of ten blasts per minute. There would be two or three thousand pulse units used per mission ...

The surviving technical drawings of the proposed ship are grandiose indeed. This was no joke. Orion was expected one day to carry thousand-ton payloads into space: ten times more than NASA's famous Saturn V moon rockets. The project was funded in earnest by the US Air Force and the Defense Advanced Research Projects Agency (DARPA). Taylor and Dyson built several small-scale test vehicles powered by conventional explosive pellets, and even launched a few mini-Orions a few hundred feet into the air from safe test sites on Point Loma, a three-mile-long spit of land guarding the entrance to San Diego harbour, and guarded itself from prying eyes by a stern military presence. The top-secret Orion project wasn't cancelled until 1965, by which time the National Aeronautics and Space Administration (NASA) was pursuing a more conventional route to the stars using chemically-powered rockets.

NASA's interplanetary space probes do sometimes carry a few grammes of plutonium in a modest, low-power device called a thermal generator, capable of delivering a few precious amps of electrical power when a spacecraft has to venture so far from the sun that solar panels become useless. The grains are clad in crash-proof canisters no larger than torch batteries, and none of them could cause any explosions. Every launch of the tiniest speck of plutonium is met by protests on the ground. The age of atomic innocence has long since passed.

PART SEVEN

'Three Quarks for Muster Mark!'

The protons and neutrons in our traditional and simplified models of the nucleus are just approximations to the truth. They, in turn, are made from yet more fundamental things.

By 1950 Feynman's life at Cornell had become a little awkward. His endlessly supportive mentor, Hans Bethe, often covered for him when he failed (as he almost always did) to fulfil his professorial obligations towards his students. He seemed childishly unwilling to tackle any paperwork or administration. No one could protect him from the sexual tensions he generated in such a small and close academic community. His flirtations with some of the wives of his colleagues threatened to spin out of control. He and Bethe both decided it was best for him to find inspiration elsewhere. After a long and refreshing trip to Brazil, he received a very warm welcome at the California Institute of Technology in 1951, a place urgently in need of a particle physics star to draw in the student crowds. Bethe was the great man at Cornell, Fermi ruled in Chicago, and Oppenheimer's doomed ghost still haunted Princeton, but here in this sunny Pasadena campus Feynman could make his own kingdom at last.

Of course his self-satisfied supremacy at Caltech did not go unchallenged for long. Another talented New Yorker, ten years younger, soon arrived to needle him. Can we be surprised to find a familiar pattern emerging in this next story? A clever but dissatisfied Viennese-born immigrant,

Isidore Gellman, comes to New York, sets up a moderately unsuccessful language school, and turns to his precociously bright son to realise his failed intellectual aspirations. Murray Gell-Mann (the hyphen came later) was born in 1929 in New York's Lower Manhattan just four weeks before the Wall Street Crash destroyed the American economy, along with the Gell-Manns' fragile language school. Isidore retreated from the world, taking comfort from studying Einstein's relativity. Murray's mother Pauline struggled along as best she could, but she resorted increasingly to a kind of dream world, somewhat like the Rose Blanche character in *A Streetcar Named Desire*. Murray and his older brother Ben had to cope on their own. They trekked to New York's great museums, one of the few resources available that didn't involve spending hard cash. Fortunately an alert teacher was intrigued by Murray's fast ability with numbers, and he was awarded a scholarship to Columbia Grammar. His parents emerged from their torpor at last, and moved to an apartment within walking distance of the school.

There was the usual fuss about Jewish quotas and customs. 'This is a Christian school run on a Christian calendar, and we want you to know that we expect you to be in class on Jewish holidays,' Murray was told. In the meantime, the boy quickly outstripped his classmates and teachers, and at just fifteen years of age he entered Yale University. His great passion was archaeology, but his father, all too well versed in the ways of the world, told him that only *rich* people went on archaeological expeditions. First, make a fortune, then do archaeology, he counselled. And so young Gell-Mann slid by default into science, and the nomadic career of a theoretical physicist. He arrived at Princeton's Institute for Advanced Study aged 21, although Oppenheimer couldn't offer him a permanent post. Then he came to Chicago University's Institute for Nuclear Studies as a relatively junior instructor, where he was glad at least to be working under Enrico Fermi, still the great man there. By 1954 Gell-Man had become an assistant professor, and his ambition was to tame what was known in those days as 'the particle zoo'.

Enrico Fermi's death from cancer that year at the age of just 53 was a shock to the entire physics community. Gell-Mann decided it was time to leave Chicago. He went to Caltech and introduced himself to Feynman. The two men locked horns for several hours and decided they quite liked each other. Still trawling for good physicists, Caltech offered Gell-Man an office on the floor above Feynman's. The next major phase of atomic discovery was about to begin; and it would be enlivened, as so often before in the history of physics, by a witty collaboration between two allies which sometimes was hard to distinguish from a sarcastic rivalry among competitors.

By now, QED could analyse the behaviour of atoms and photons, and the exchanges between matter and electro-magnetic energy, to such a degree of accuracy that the difference wasn't worth a damn. Still, that wasn't the same as *understanding* all the constituents of an atom, least of all its nucleus. As the range of new particles and sub-particles emerging from the latest accelerator tests threatened to multiply out of control, physicists were stuck with the problem of explaining them all. And still a number of deeper mysteries persisted. What was the 'strong nuclear force' that held the nucleus together, and which could be calculated to a nicety, yet which had no physical explanation? Come to that, where did the mass of particles come from? William Pollard, director of the Oak Ridge Institute, expressed the continuing frustrations of his trade: 'Whatever it is that explains a property cannot itself possess that property ... The ultimate solution of the atomic quest will be to arrive at a really elementary component which accounts completely for all the diversity and behaviour of matter, yet does not itself possess any observable property whatever.' Pollard feared that we could go on dissecting the subatomic realm ad infinitum without ever touching the bottom.

In 1961 Gell-Mann and an Israeli collaborator, Yuval

Ne'eman, came up with a scheme for classifying particles into a new and orderly arrangement of families. With typical linguistic inventiveness, Gell-Mann christened the scheme the Eightfold Way, in honour of Buddha's Eightfold Path to Enlightenment. It grouped protons and neutrons (any particles affected by the strong nuclear force) into subgroups or 'multiplets', each one governed by the contributions of even more fundamental particles. He later called these basic bits of matter 'quarks', adopting the fanciful term from James Joyce's novel *Finnegans Wake*. The electron, meanwhile, retained its 'fundamental' status.

Gell-Mann's theory initially required different kinds or 'flavours' of quark. A proton, for instance, seemed to consist of an 'up' and two 'down' quarks nearly, though not quite, equal in mass. It took all three, in this lop-sided arrangement, to generate the overall unit charge of the proton, and this immediately upset some of Gell-Mann's critics. How could a positive charge be divided into *fractions* of a charge in each quark? Then there was his explanation for why quarks couldn't be liberated in accelerator experiments. The force of attraction between them, mediated, apparently, by carrier particles called 'gluons', did something very peculiar. Like an immensely powerful rubber band pulled tight, it grew stronger with distance, so that whenever an experimenter tried to separate quarks by force, their bonds tightened to match the challenge. However, when left alone inside the bounds of a proton, those quarks apparently drifted about quite peaceably, as though their rubber band had relaxed.

At first, in Gell-Mann's own words, 'Quarks went down like a lead balloon.' How could quarks exist inside a proton or a neutron without violating Pauli's exclusion principle, the law that said no two particles could occupy the same space? His confidence wavered: 'Even I thought the idea of unobservable fractionally charged particles was crank.' One critic, Sheldon Glashow, was disturbed by the apparent difficulty of liberating quarks in any collision experiment.

This seemed like a cheap get-out clause. 'You can't even get them out with a quarkscrew', he complained. Even when Gell-Mann stepped up to the podium to receive his Nobel Prize in 1969 'for his contributions and discoveries concerning the classification of elementary particles and their interactions', he was still hedging his bets: 'The quark is just a notion so far. It is a useful notion, but actual quarks may not exist at all.' Some people muttered that he was playing both ends against the middle. If quarks turned out not to be real, he could say he never claimed they were. If quarks *were* eventually verified, he would remind everyone that he discovered them.

He seemed to be on the right track. Experiments at the Stanford Linear Accelerator, in which extremely high-energy electrons bombarded protons, had liberated a shower of sub-particles. Feynman picked up on these experiments and worked for a while on his own theory, in which a proton contained 'partons'. Gell-Mann was mightily displeased. 'The whole idea of saying that they weren't quarks but some new thing called "put-ons" seemed to me an insult.' The other thing that really began to irritate him about Feynman was the success that his sometime friend and sparring partner was beginning to have as a media star. Caltech never imagined it could tame Feynman into a responsible lecturer for his students, so they came up with another idea instead. Would he like to give an introductory course to that most disregarded section of the student body, the freshers? Feynman asked, had any such senior physicist as himself ever taken on such a lowly role before? When assured that he would be the first big-name physicist to accept such an apparently humble job, he leapt at the chance. His series of talks, delivered between 1961 and 1963, were among the most eloquent and inspirational discussions of his craft that had ever been made, even if they were not quite suited to his undergraduate audience. Acolytes carefully transcribed his every word, and when published in book form they were a great success. Gell-Mann disparaged them as 'Dick's joke books', while he struggled somewhat resentfully to overcome his own terrible writer's block.

On the plus side, his quark theory prevailed. More results from the accelerators required him to devise three more kinds of quark: 'top', 'bottom' and 'charmed'. The six-strong family also needed properties defining their three kinds of fractional charge, which Gell-Mann decided to call 'colour'. When he chose 'red', 'white' and 'blue', his enemies lost no time reminding this proudly pedantic linguist that he'd been slipshod in his thinking. White is not a primary colour. Today – for the quark theory still prospers – colours are divided into red, *green* and blue. They have absolutely no relation to colours as we would know them. They are shorthand terms for mathematical values, but we owe to Gell-Mann the quirky nomenclature of the quarks. As the showers of particles produced by the accelerator collisions spontaneously split into two, then finally three generations of particles, so the new quark theory, quantum chromo-dynamics (QCD instead of the old QED) repaired and maintained its predictive powers.

Feynman may have lost his seat at the High Table of physics to Gell-Mann and his allies, yet he had at last found happiness in his domestic life. In 1960 he married a sparky and intelligent young woman from Yorkshire, Gwyneth Howard. Alone among all the women he had known since Arline, Gwyneth fell in love with him for what he was, rather than for what she hoped he might become. Throughout the 1960s and 1970s, Feynman continued to achieve what others could not. He became adept at telling the story of scientific inquiry to an ever-larger audience, and made his difficult and abstruse trade sound like a lot of fun. As usual, Gell-Mann couldn't decide if Feynman was his good friend or his annoying competitor. When Feynman died of cancer in February 1988, just a few weeks shy of what would have been his 70th birthday, Gell-Mann upset a number of people with a grudging passage in his eulogy. There had been, he said, 'a well-known aspect of Richard's style. He surrounded himself with a cloud of myth, and he spent a great deal of time and energy generating anecdotes about himself.' And it was quite true. But so what?

Quite apart from Feynman's ebullient character as a physicist, his role in the making of the atom bomb and his elegant diagrams, he caught the public attention one last time before his death, when he investigated one of America's most distressing rocket accidents. The NASA space shuttle *Challenger* was destroyed on 28 January 1986, less than two minutes into its flight. One of its solid rocket boosters sprang a leak. Freezing weather had affected the rubber rings that sealed the various sections together. A jet of flame escaping from the faulty seal burnt into the liquid fuel tank and the entire ship blew up just 70 seconds after lift-off. Six astronauts and one civilian teacher, Christa McAuliffe, were killed. The shuttle fleet was grounded for an intense and emotionally gruelling investigation. Feynman was asked to join the board of inquiry looking into the disaster.

At first he wasn't sure if he was ready to do this, but Gwyneth was keen that he should have something interesting to do, to divert him from the cancer that was killing him. She persuaded him that he could make much more trouble than anyone else on the board. And he proved her right, cutting his own path through the verbiage and applying some simple logic to what he called the 'fuzzdazzle' of NASA technical presentations. With the TV cameras running, he dipped a piece of rubber into a glass of iced water and showed how it hardened when cold. 'Do you suppose this might have some relevance to our problem?' he asked, knowing very well that it did. He had created a vivid demonstration of the notorious 'O' ring flaw that had destroyed the shuttle during its wintry launch. But Feynman wanted to probe further. 'If NASA was slipshod about the leaking rubber seals on the solid rockets, what would we find if we looked at the liquid-fuelled engines and all the other parts that make up a shuttle?'

Feynman was told that the inquiry was not briefed to look

at the main engines, because no problems had been reported. So he made unauthorised trips to NASA facilities where he could speak to ground-floor engineers in private. He wrote later: 'I had the definite impression that senior managers were allowing errors that the shuttle wasn't designed to cope with, while junior engineers were screaming for help and being ignored.' He had identified a serious human problem that haunts NASA to this day. In the inquiry's final report he made a plea for greater realism: 'For a successful technology, reality must take precedence over public relations, because Nature cannot be fooled.' This last bravura act of his brilliant career tells us that the logic, the clarity of mind and the intellectual honesty required to probe the tiniest secrets of the atom can apply equally well to problems at our more familiar human scale.

The nano-world

In 1959, Feynman made a prediction that sounded pretty wild at the time: 'The principles of physics do not speak against the possibility of manoeuvring things atom by atom. In principle, it should be possible to synthesise any chemical substance. How? Put the atoms where the chemist says, and so you make the substance.' Three decades later, Eric Drexler popularised the idea of achieving 'thorough control of the structure of matter at the molecular level. It entails the ability to build molecular systems with atom-by-atom precision, yielding a variety of nanomachines.'

In 1981, Gerd Binnig and Heinrich Rohrer of the IBM Zurich Research Laboratory invented the Scanning Tunnelling Electron Microscope (STEM). A super-fine tungsten probe – its tip may contain just a single atom – is moved across the surface of a sample. The movements are controlled by nothing so crude as gears and wheels, but by piezo-electric quartz crystals, as originally investigated by Pierre and Jacques Curie, that change shape according to the strength of an electric current applied to them. As the tungsten probe scans its target, a measurable charge flows between the tungsten atoms and those in the sample, because electrons bridge the gap via a process called quantum tunnelling (Gamow's contribution). The piezo-electric crystals also act as a feedback device, measuring how much counter-force is required to keep the probe at a constant distance from its target. Atoms can be magnified 100 million times and visualised in the form of contour maps on a TV screen. In 1990, Don Eigler and Erhard Schweizer of the IBM Almaden Research Center in California spelt out the letters IBM with 35 xenon atoms arranged on a nickel surface, using the tip of a STEM to manoeuvre individual atoms.

Manufacturing – say, of a fine clockwork watch – usually involves taking large chunks of metal and whittling them down into the required cogwheels and springs. It's an inefficient procedure, because most of the metal ends up on the workshop floor as discarded shavings. In addition, huge amounts of energy are needed to extract pure metal from crude ores in the first place, and then to melt the metal into bars, plates or rods that can be yet more finely cut up and shaped by the watchmakers. Nanotechnology turns all this processing on its head, building upwards from the smallest scales by molecular accretion. Biology functions like this all the time. We are now entering an age when some of the distinctions between biology and engineering are starting to blur. We will grow our machines, rather than building them. We might perhaps become capable of reshaping our world atom by atom, even though we are still not quite sure what atoms are.

Today, in what's known as the 'standard model' of theoretical physics, it's generally accepted that all elementary particles of matter belong to a family known as fermions. Another class of particles also exists, called bosons. These are force carriers. Today there are believed to be just two species of fermions: quarks and leptons. The latter is the name for all particles that do not feel the strong nuclear force by exchanging gluons; that is, all elementary matter particles that are *not* quarks! Leptons include the electron and its two heavier relatives, the muon and the tau, as well as three types of neutrino (a very light and elusive particle), known by their association with the other three leptons. Thus there is the electron-neutrino, the muon-neutrino and the tau-neutrino ... And there's worse to come. Physicists are on the look-out for yet more particles. For instance, the Higgs boson is theoretically responsible for endowing other particles with their appropriate masses; and the long-anticipated graviton should be the boson that conveys the force of gravity. And this is not to mention what are called 'supersymmetric' partner particles for all of the above ...

Our quest to understand the atom still has a long way to go.

PART EIGHT

Ylem

Ν

Unveiling the forces inside the atom gave us clues about the origins of the entire universe. The world of the extremely small became inextricably linked with the cosmos at the grandest scales.

In 1905, a five-year-old English girl, Cecilia Payne, saw a meteor shooting across the sky and resolved at once to become an astronomer. Fourteen years later, as a student at Newnham College, Cambridge, she attended a lecture by Arthur Eddington in which he explained how his solar eclipse observations had proved Einstein's prediction that a massive object such as the sun should bend the light from distant stars. She tried to get access to the university's telescope, only to find its guardian rushing off in consternation to warn his superiors: 'There's a *woman* out there, asking questions!' Payne's career would be dogged by the most absurd sexism and resistance, yet she somehow managed to make one of the most important discoveries in all of astronomy.

Eddington took her on as a tutorial student. A shy and formal man, he never met with Payne unless his elderly unmarried sister was also in the room to act as chaperone. The young woman and the prim gentleman struck up a certain friendship. He opened her mind to the possibilities of theoretical physics, at one point even arguing that the human mind should be capable of inferring the existence of stars and planets even if the earth's skies were perpetually clouded and no telescopes had ever been built. Despite his support, Payne's career could not prosper at Cambridge. She went to Harvard, hoping for better luck, only to discover that the best work that a woman could get was as a 'computer'. By 1923 Harvard maintained a small army of pencil-and-paper human calculators, almost invariably female, who crunched the numbers supplied by their more privileged male colleagues until their eyes smarted and their shoulders ached. Their results were essential for confirming or refuting the grander theoretical ideas of the men, but the chances for a woman to make any original contributions in research were limited. Payne was appalled but undaunted. She began analysing the complicated spectra of light from stars, and found that the entire astronomical community was misreading them.

The prevailing theory was that stars like our sun consisted of around 60 per cent iron. Payne believed that certain ambiguities in the way that the spectra were read meant that the entire set of results should be skewed significantly towards the lighter elements. In a somewhat brave Ph.D paper, 'Stellar Atmospheres', she concluded that the sun was over 90 per cent hydrogen, with most of the rest consisting of helium. Contrary to the beliefs of the old guard, iron was a *minimal* component. Payne's tutors were outraged, and when she kept to her guns and formally delivered her paper, she was forced to include the weasel phrase: 'The enormous abundance [of hydrogen] is almost certainly not real.' She struggled also against the sexism of her times: 'One serious obstacle existed: there was no advanced degree in astronomy, and I should have to be accepted as a candidate by the Department of Physics. The redoubtable Chairman of that department refused to accept a woman candidate.' Harvard decided to found an Astronomy department so that it could award the Ph.D. Even so, no one but Payne believed her results.

Other astronomers gradually recognised that Payne was right. Her work was eventually heralded as one of the most brilliant contributions in astronomy. Now the path was clear to discover how our sun, and the other stars, generate such vast amounts of energy over so many billions of years without burning themselves out. It is a *nuclear* process. Hans Bethe and others looked at what happens when hydrogen is compressed in the heart of the sun. The unimaginable gravitational pressures, four trillion pounds per square inch, squeeze hydrogen nuclei together until helium is formed, in a process known as 'fusion'. But the mass of helium that emerges from this process is 0.7 per cent less than the mass of hydrogen that goes into it. That 'missing' mass is liberated as energy. The sun converts four million tons of hydrogen into energy every second, yet the amount of available hydrogen is still so great that it should be able to sustain this process for another 5,000 million years ... The sun was no longer quite so mysterious as it had been. On the other hand, no one yet knew where it, or anything else in the universe, had *come* from. Everything was made from atoms. What had made the atoms?

Black Sea boat to freedom

George Gamow, the colourful Russian whose motorcycle exploits had so disturbed the tranquillity of the Cavendish Laboratory's social get-togethers, returned to Russia in 1931, where he accepted the post of physics professor at the University of Leningrad. He was frustrated when the Soviet authorities barred him from foreign travel, and at one point he tried unsuccessfully to reach Turkey by escaping across the Black Sea by boat. On being recaptured, he somehow persuaded the authorities that he had just been testing the boat's performance! At last he and his wife Lynbov were given permits to attend the 1933 Solvay Conference in Belgium, and they seized their chance to quit the Soviet Union for ever. Gamow spent the rest of that year visiting various institutions all over Europe, catching up with old friends and hearing the latest theories. Then he settled for the next two decades at George Washington University in Washington, DC.

By now he was intrigued by the stars. Astronomers knew from spectroscopic analyses of thousands of stars (and especially once Cecilia Payne had corrected their mis-interpretations) that the elements were distributed in more or less similar proportions all across the observable universe. Why should it be, Gamow wondered, that hydrogen and helium accounted for almost all matter in the universe, while oxygen, carbon, nitrogen and the other heavier elements accounted for barely 2 per cent? Where did the comparatively rare heavier elements come from?

In 1929, the American astronomer Edwin Hubble demonstrated that all the galaxies in the observable universe are flying apart from each other. The further away a galaxy is, the faster it recedes. The light from distant galaxies stretches into longer wavelengths at the red end of the spectrum, just as sound waves from a speeding train lengthen and lower their pitch when the train hurtles past. By 1931, a scientific-ally-trained Belgian-born Catholic priest, Georges-Henri Lemaître, had electrified, not to say annoyed, many astronomers with an extraordinary idea, simply by imagining Hubble's expansion time-reversed – running the film backwards, so to speak: 'We could conceive the beginning of the universe in the form of a unique atom, the atomic weight of which is the total mass of the universe. This highly unstable atom would divide in smaller and smaller atoms by a kind of super-radioactive process.' With a brilliant leap of imagination he had described what we all know today as the Big Bang theory.

Arthur Eddington and many other astronomers at first disliked Lemaître's idea. Some were disturbed by this implication of a biblical 'moment of Creation', while others were simply too used to the idea of an infinite cosmic timespan with no beginning and no end. But Gamow was inspired. In 1948 he, too, suggested that the universe must once have been a tiny fraction of its current size, much denser, and fantastically hot. He believed that it was impossible for atoms to exist in that concentrated inferno, because the kinetic heat energy was so intense that it was stronger than any possible nuclear bond inside an atom. He surmised that the universe began as a super-compressed raging inferno of protons, neutrons and electrons, which he and his collaborators called the *ylem*, from an ancient Greek word meaning 'before the beginning of time'. The internal pressures of the ylem caused it to expand outwards extremely rapidly. After a few minutes the ylem cooled down sufficiently to allow the strong nuclear force to bind neutrons and protons together into deuterium nuclei (recall that deuterium is the heavy form of hydrogen, with a neutron included in the nucleus). Then there was just enough time for some of the deuterium to fuse into helium before the temperatures and pressures of the ylem became too diffuse to allow more fusions. Gamow drew short of trying to explain where the ylem came from in the first place; nor could he explain how any of the elements heavier than hydrogen or helium had been manufactured. But the seed of a major new idea was taking root.

Gamow's principal collaborator in this work was one of his students at George Washington University, Ralph Alpher. When Gamow told him to start thinking about the origin of elements, he couldn't have found a young man more primed and ready to upset the conventional ideas about Creation. At just eleven years of age, Alpher had argued with his rabbi Hebrew teacher about the meaning of Genesis: 'It got hot and heavy. I'd bring in quotations, with references, and he didn't want to have any part of it.' He was very meticulous with his maths, and this was fortunate because Gamow was sometimes quite a sloppy calculator, more concerned with the dramatic flourishes than the fine detail of a theory. Science needs both these kinds of talent. Between them, they wrote a letter for the April 1948 edition of the *Physical Review*. Just before sending it off, Gamow asked Hans Bethe to sign his name to the paper.

It was known forever afterwards as the 'Alpher Bethe Gamow' theory ...

In collaboration with Robert Herman at Johns Hopkins University, Alpher then made one of the most important predictions in the history of cosmology. If the ylem had been as hot and energetic as he and Gamow suggested, then the remnants of that heat should still suffuse the universe today, as an extremely faint but nevertheless detectable afterglow of electromagnetic radiation. The temperature of the vast and supposedly frigid-cold wastelands between galaxies should not be quite zero after all. Alpher couldn't find any radio experts to help back up his theory. Radio astronomy was not well developed at that time.

Meanwhile, an argumentative Yorkshireman found the ylem a highly distasteful idea, and lost no time saying so. He couldn't have been more wrong about the origins of the universe, but his explanation for where the elements came from would turn out to be sounder than Gamow's.

The Yorkshireman

The confrontational and gruff-spoken Yorkshire-born cosmologist Fred Hoyle lost his respect for authority as early as primary school. A teacher told his class that a certain type of flower has five petals. The next day, Fred produced a flower of the same kind with six petals and asked the teacher why she'd said it had five. The teacher clouted his ear. Fred left school at once, walked back home and told his mother he refused to go back to such an unfair place. Battle was joined with the education authorities, who eventually allowed Fred to attend a better school. As a young man he displayed an astonishing talent for mathematics, graduating from Cambridge University with the highest marks in his year. This won him a coveted position as a research student at the Cavendish Laboratory, then still at the height of its prestige. In 1938 he persuaded Paul Dirac to be his supervisor, on the strict condition that the taciturn Dirac wouldn't have to talk to him too often. Just one year under Dirac's silent tutelage enabled Hoyle to produce two brilliantly successful papers in quantum electrodynamics.

In the early 1940s he contributed to the new and largely British-led development of radar, although military work did not appeal to him, and he quit as soon as he could once the war was over. A secret trip to America in 1944 brought him into contact with some of the atomic bomb theorists. An astronomer called Walter Baade set Hoyle's mind racing when he commented: 'Maybe a star is like a nuclear weapon?' Hoyle wondered what might happen if the nuclear chain reactions were *constantly sustained* inside a star, instead of merely dissipating after a few spectacular moments in a bomb explosion.

An Anglo-American physicist, Geoffrey Burbidge, and his astrophysicist wife Margaret Burbidge, took up the cause and worked with Hoyle on a mammoth calculation that took half a decade to complete. Margaret found specific astronomical evidence that nuclear reactions occur in stars, while her husband calculated the proportions of chemical elements discernible in their light spectra. There was growing evidence that many elements apart from hydrogen and helium existed inside certain stars. But how were they made? Certainly not in Alpher and Gamow's ylem, Hoyle thought. Instead he decided that everything hinged on the stellar creation of carbon-12 (i.e. carbon with six neutrons and six protons). In the spring of 1953, during another American visit, Hoyle burst into the office of his good friend Willy Fowler, a nuclear physicist at Caltech with an expert knowledge of carbon atoms. 'I exist, and I'm made out of carbon!' Hoyle exclaimed. 'Therefore the carbon-12 nucleus *must* possess an energy level at 7.65 megaelectronvolts (MeV).' This was his technical shorthand for an amazing new idea he'd come up with.

Hoyle believed that the story of nuclear reactions inside a star could be far richer than just fusing hydrogen nuclei into helium nuclei – and especially when it came to stars substantially larger and hotter than our own sun, which explode violently at the ends of their lives, briefly creating yet greater temperatures and pressures. In these stars, two helium nuclei can come together to make a nucleus of beryllium, but only in a form that is highly unstable, splitting apart after just one billion billionth of a second. If only Hoyle could figure out a way of fusing helium into something which *didn't* blow apart after a vanishingly small instant of time ... Otherwise his revolutionary idea – that almost all the elements we know of apart from hydrogen are created inside stars – would come to nothing. So he asked: what would happen if *three* helium nuclei collided? This would be only a rare event, but by the sheer laws of probability it had to happen sometimes. Hoyle claimed that the product of these collisions would be carbon-12. His prediction, conveyed excitedly to Fowler, was that a hitherto unsuspected energy level should be detectable in carbon-12 nuclei. He proposed a startlingly exact figure of 7.65 million electron volts. If he was right, carbon-12 was manufactured inside stars, paving the way for the 'nucleosynthesis' of all the heavier elements too.

Everything depended on the combined quantum behaviour of those three helium nuclei. What if, just for the briefest moment, they created a resonance, like three very finely-tuned violins all reinforcing the same note? Hoyle calculated that this resonance, although short-lived, should allow the creation of carbon nuclei, which would come into existence at an energy level of 7.65 million electron volts. Now all he needed was proof that carbon was actually capable of exhibiting that level, which had never before been observed.

Although Fowler would later tell people that his first impression of Hoyle that day was of someone who had lost his mind, he listened patiently, then persuaded a team of experimenters at Caltech's Kellog Radiation Laboratory to put the theory to the test by accelerating carbon-12 nuclei to the correct energies. For ten days, Hoyle tried to contain his impatience. At last the team emerged from the lab and shook his hand. He was right, and the energy levels he'd predicted were so close to the experimental results that the difference scarcely mattered. What surprised Fowler, even more than the unnerving accuracy of Hoyle's calculations, was the style of his approach when he'd first come storming into his office. Hoyle had used what's called an 'anthropic' argument. Life couldn't arise without carbon, so there had to be a mechanism in the laws of nature to generate it in the right way, and with that special and extremely fine-tuned reson-ance. Otherwise none of the other elements could be formed, and life would not exist. Fowler thought this sounded like a quasi-religious and unscientific idea. He and Hoyle struck up a long friendship nevertheless.

Margaret and Geoffrey Burbidge, Fowler and Hoyle, the so-called 'B2FH' team, set to work accounting for all the elements in the periodic table. Their huge compendium of results and mathematical proofs, running to more than 100 pages, was published in 1957. It was an absolutely epochal moment in science. We are made from atoms created in stars, and now we understand something of the process:

A certain kind of massive star, at least ten times bigger than our sun, 'burns' extremely fast, because its huge gravity field compresses the core, raising the temperature to 100 million degrees and more. The star's hydrogen is converted to helium in only 100 million years – just 1 per cent of our own sun's lifespan. As the energy given out by the nuclear fusions is reduced, so the internal pressure drops, and the outer layers of the star collapse as gravity forces the void to be filled. The core temperatures rise again, and now it's time for the newly-minted helium to fuse. Hoyle showed beyond any doubt that helium could transmute into carbon-12, paving the way for heavier elements yet to come. The interior pressure falls once again as the latest load of fusion fuel is used up, the star collapses again, and the nuclear furnace is reanimated as the pressures and temperatures revive. And now some of the carbon-12 (with six protons) fuses with helium to make oxygen. Then comes silicon and, eventually, iron. A supernova explosion at the end of the star's life hastens the reaction while extremely rapidly creating elements *heavier* than iron. Then the force of the explosion scatters these products far into space, where they eventually coalesce inside other second-generation stars and planets.

With the exception of hydrogen, every atom inside our bodies was created in the heart of a long-vanished super-massive star. We are the products of an ancient supernova; or more likely, several supernovae whose products merged gently over billions of years to create vast clouds of inter-stellar dust, from which our sun and our earth eventually coalesced some 5 billion years ago. Yet more supernova explosions must have sent shockwaves through the cloud, causing random concentrations of dust, around which the first microscopic seeds of new suns such as ours were born.

Fowler and other scientists were eventually honoured with Nobel citations for this discovery. But Hoyle was not, even though he was without doubt its leading champion. His brilliance and farsightedness were matched by an argumentative streak, and by a fondness for certain other theories that were way beyond anything his colleagues were prepared to agree with. Just as Fowler suspected, Hoyle thought the universe was *purposeful*, specifically suited to the emergence of life. This was a stretch too far for most of his colleagues, let alone his enemies.

One last stand

Hoyle's other raging passion had some of the character of Einstein's futile last stand against the quantum mechanics. While the scientific community at large was drawn inexor-ably towards the idea of an explosive instant of Creation, somewhat along the lines that George Gamow had described, Hoyle would have none of it. Along with the mathematician and cosmologist Hermann Bondi, and the astronomer and geoscientist Thomas Gold, he argued that the universe had no beginning and will have no end. It is eternal. This concept, known as the 'steady state', won few allies because it seemed too much like yet another religious idea. As Hoyle put it: 'For every volume of space the size of a one-pint milk bottle, about one atom is created every thousand million years.' This was enough, in his scheme, to keep the universe in its steady state by replenishing the matter dissipated by the flying-apart of the galaxies, eventually allowing new galaxies to form in the thinned-out regions. Critics argued that Hoyle couldn't account for how atoms could spontaneously appear out of the vacuum. He replied that the Big Bang's sudden emergence out of nothing was even more inexplicable. In fact that term, Big Bang, was his derisory coinage.

Proof of the Bang?

Recall that Gamow's young colleague Alpher had predicted that there should be a detectable echo of Big Bang, a fractional heat energy suffusing all of space. In 1965 two radio astronomers, Arno Penzias and Robert Wilson of Bell Laboratories in New Jersey, were testing a sensitive horn antenna as part of a project to try to identify certain frequencies of radio noise that could interfere with communications satellites. The first thing they needed to do was calibrate their antenna against a zero-strength signal by pointing it at an empty region of sky, a spot where the density of stars or distant galaxies should have been so low that no radio noise would be detected. They were vexed that a constant low-level hiss of microwave background noise refused to go away. They checked and rechecked all their gear, and even climbed inside the dish of their radio antenna to evict a pair of pigeons. The pigeons came back – they were homing pigeons, after all – so somewhat reluctantly they persuaded a local pigeon expert to *shoot* them. Then they went back inside the dish yet again to clean off the old pigeon shit, just in case this 'white dialectric material' was somehow generating radio interference. They tried everything, but the irritating background hiss just wouldn't stop. They moved the antenna around the sky, looking for another quiet spot. There were no quiet spots. Penzias and Wilson realised that the faint radio 'interference' was coming from the entire sky. It was the echo of Big Bang.

A rival group at Princeton just a few miles away had actually been looking for that radiation. They took the news quite well. Gamow and Alpher, however, were furious that they had not been credited for their brilliant prediction of this epic discovery. The slight was unintentional. Wilson and Penzias had barely even known about Big Bang theory, let alone the work of its principal theorists from two decades past. Aside from the human wrangling, a key moment had been reached. The sub-microscopic atom was connected in a seamless chain of reasoning to the surrounding vastness of the universe.

We now know we were made from atoms, and we know when and where and how they were created. We even know that time and space were born in a pinprick of matter smaller than an atom, yet infinitely dense.

Now the *real* questions start.

PART NINE

New Frontiers

Modern physics plunges us into speculations that sound like the craziest science fiction, yet they are founded on very serious theories. Can we ever discover the ultimate nature of reality?

In the last two decades, atomic physics has reached a new impasse just as severe as the renormalisation-of-infinities problem that assailed theorists in the 1940s. This time the challenge has been to connect the hyper-small world of atomic forces with gravity, the dominant influence over large-scale entities like stars and galaxies. Gravity's effect on individual atoms is not yet understood. Yet the universe is made out of atoms, and obviously the universe is shaped primarily by gravity. The search is afoot for the 'graviton', the particle that (perhaps) conveys the force of gravity. So far, we haven't found it – and it may be a complete theoretical fabrication.

The graviton problem is simplified if we think of particles as just the surface manifestations of something deeper. Many scientists believe that all particles of matter, and all force-carrying particles too, are generated by extremely tiny vibrating 'strings' of energy. One mode of vibration, or 'note', makes a string behave as an electron, another as a photon, and so on. Best of all, string theory includes a vibration mode to deliver the graviton. At last, gravity can be described subatomically as well as at the cosmic scale.

All of physics might soon be tamed so that it appears akin to harmonious music from a single stringed instrument, rather than a bewildering cacophony of noises from an argumentative orchestra. The hope is that a string-based Theory of Everything will bind all the laws of nature into one equation. It will explain the entire universe as surely as a single atom. Each atom, in fact, will just be a phrase in the string vibration, a temporary set of harmonics, and not a discrete 'thing' as we previously imagined. On the other hand, strings make some sense of the wave–particle duality that has vexed us for so long. The universe as we see it is not a collection of particles. It's a *performance* derived from myriad vibrations. Our scientific instruments are tuned to pick up particular modes of vibration, and no single machine can hear them all ...

One snag. Strings are much smaller than particles, so we have no obvious prospect of detecting them (proving their existence) with even the most advanced instruments that study particles. Another difficulty. While the mathematics of string theory seem robust, they require outlandish assumptions. We experience the world in four dimensions: one in time, and three in space. Strings supposedly exist in ten or eleven dimensions, and perhaps even more, depending on the specific theory. The 'hidden' dimensions are wrapped up in invisibly tiny bundles, associated with the strings, called Calabi-Yau manifolds. This may be a convenient method of sweeping the excess dimensions under the carpet, where no one has to worry about them in practical terms. They are there, the theory says, but so tightly squeezed together that we have no way of observing them. Right now, string theory is under vigorous attack from experimenters who refuse to believe in something they can't test in the lab. Our investigation of the subatomic realm continues to plunge us into a world of perplexity.

Multiple realities

One interpretation of the wave–particle duality that we've encountered time and again in this book is known as the 'collapse of the wave function'. When we're not looking at it, a particle is a wave of possible locations and interactions until the moment when an observation is made. Instantly the wave collapses to a single point, which we observe as a particle impinging on our instrument. One problem with this is related to the EPR paradox. If the wave–particle has been travelling across space for some while, and the wave has spread out to become extremely large, then how does the entire wave 'know' to collapse to a single point as soon as we make a measurement? Different arcs of the wave, separated in space, would need to communicate with each other faster than light in order for the entire wave's collapse to be instantaneous.

The strangest possibility of all is this. All the potential outcomes that the wave–particle *could* have delivered from its collapse are still 'out there' when we measure a particular particle. Twenty years ago, the idea of multiple universes seemed a fantasy. Now it threatens to become scientifically necessary. The mathematics of quantum theory suggests that subatomic wave–particles shimmer with *possible* states, ghostly clouds of probabilities and potentialities, living-and-dead cats. When we record specific particle events and qualities, we build up a history, through time, of 'what has happened' in our world. Observations shape our reality, but all the different particle events that we *might* have observed are perhaps no less real. It's just that they happen in parallel universes superposed on top of ours. With every passing moment, alternative realities accumulate in the quantum 'multiverse'. Somewhere, another you exists that has not chosen to read this book.

Richard Feynman told a characteristically colourful story in which he goes to a Las Vegas casino, and he's about to put a large bet on the number 22 on the roulette table. The pretty girl next to him spots someone she knows out of the corner of her eye and spills her drink. Feynman's attention wanders, and he is delayed by a fateful fraction of a second and fails to place his bet in time. Number 22 comes up a winner. He could have been rich, but instead leaves the casino a ruined man. 'I can see that the whole course of the universe for me has to hang on the fact that some little photon hit the nerve ends of her retina. Thus the whole universe bifurcates [splits into two] with every atomic event.'

Meanwhile, a brilliant British physicist, David Deutsch, says that scientists have proof of multiple universes staring them in the face, yet they refuse to see it. In the two-slit experiment, a single electron does not interfere with itself after all. Instead, it is interfered with by its counterparts in parallel universes. Particles that appear to behave mysteriously and unpredictably are actually being nudged about in a perfectly classical and deterministic way by other particles that we cannot see. It might sound a crazy idea, but it would yield a classical universe once more! Albeit one with an extremely large number of parallel dimensions superposed on top of each other.

The evolving cosmos

Quite apart from this quantum (or maybe even classical) shimmer of multiple universes, there may be another mechanism at work in the cosmos which produces them via black holes.

Physicists have been able to backtrack the story of cosmic creation to within a tiny fraction of a second of Big Bang, using the laws of physics as we know them today. But where do those laws come from? For all their failings, mathematical models of the cosmos, from stars and galaxies down to the orbits of simple asteroids, have proved incredibly effective. But why *should* numbers and equations be so useful? Albert Einstein often commented: 'It is not the laws of the universe that are remarkable, but the fact that there are any laws at all that we can comprehend.' More recently, Stephen Hawking has asked: 'What is it that breathes fire into the equations and makes a universe for them to describe? Was it all just a lucky chance? That would seem a counsel of despair, a negation of our hopes of understanding the underlying order of the universe.'

Hawking certainly does not suggest God as his solution, but he does demonstrate that we haven't yet found a complete explanation for the universe's existence. We can describe the sequence of events that led to it being here, but we are still trying to figure out how and why it all happened. This is where we explore the Anthropic Principle, as favoured by Fred Hoyle. In its 'weak' form it's simply a philosophical query. In its 'strong' form it speculates that the cosmos might be, as Hoyle once called it, 'a put up job'. The 'weak' anthropic principle works like this: the universe seems suspiciously well-suited for life. We think like this because we happen to exist within that universe. Had the universe turned out to be unsuitable, we wouldn't be here now to discuss the problem. Okay, the weak version doesn't really get us very far. The 'strong' anthropic principle is much more interesting. It examines the four fundamental forces of nature, and looks at what might have happened if they had turned out differently.

Everything in our universe can be accounted for by just four natural forces: gravity, electromagnetism, the strong nuclear force which holds the nucleus of atoms together, and the weak nuclear force which allows radioactive atoms to decay. What would happen if any of these forces were to change, even slightly?

Gravity is such a weak force at small scales that it scarcely counts. However, if the exact strength of gravity weakened by the tiniest degree, the large-scale implications would be dramatic. Galaxies would disintegrate. Suns would die prematurely. The formation of new suns from drifting galactic dust clouds would be impossible. We wouldn't be here, because the cosmos couldn't have produced stars such as the one that warms our earth. On the other hand, if gravity were a little stronger, then the rate of collisions between stars would be so great that a typical solar system like ours wouldn't survive long enough to produce stable planets. We wouldn't have had a chance to evolve.

What about electromagnetism? If the balance of the electromagnetic force altered in any way, all known chemistry would disintegrate. The same applies to the strong force. As for the weak force, it's tempting to think that we could do without radioactivity; but without it, the earth would not have been able to maintain its heat from its core, and would have cooled down billions of years ago. We would not have evolved. These examples are gross simplifications. Any tiny change in the relationship between any of the four forces would produce not just the minor inconveniences described above, but the dissolution of the cosmos as we know it.

The four forces that make the universe possible can be found within a single atom. Describe the behaviour of an atom in terms of those forces, and the characteristics of stars, planets and galaxies follows with absolute inevitability. The same applies to molecular biochemistry and life, because we're all made of atoms. Hawking's biographer Kitty Ferguson has observed: 'If something in the cosmos isn't a product of the four forces, then it hasn't happened.' That includes us humans.

The strong anthropic principle presents us with the following very exciting idea: at the same instant that our universe was created by the Big Bang, the laws of physics as we know them today must also have been created, because those laws are a property, a characteristic, that we find in this universe. *But they didn't have to be the way they are.* There is no law-about-laws which dictates that the four fundamental forces had to come out the way they did. Most physicists agree that the Big Bang could have produced forces and physical laws very different from the ones we actually find out there. In fact the possible varieties of unstable balances between the forces are infinite. Any of these 'poor' balances would have produced a universe incapable of maintaining order and creating life. The odds against our stable, life-friendly universe emerging from the Big Bang were vastly greater than those against winning the National Lottery grand prize.

So how did Big Bang apparently generate a 'good-quality' universe first time out of the hat, against such long odds? The strong anthropic principle (in one of its many versions) seems to say that we might be able to prove the existence of God by virtue of this incredible statistical fluke. The likelihood of the universe coming out badly was so high that only a deliberate design could have biased the betting in its favour. Physicists have a nickname for the mathematical values of the four forces we discussed earlier. They call them the 'hand-set values' as an acknowledgement of how unlikely it should have been for those values to balance out so neatly.

In fact the anthropic principle doesn't specifically promote the idea of a Grand Designer. It's more about learning to ask the right kind of questions about our universe. Those scientists who agree that the fluke of the Big Bang has to be explained, but who do *not* believe that a designer was involved, have come up with some clever theories of their own. One of these calls on the mind-boggling properties of black holes.

When some very large stars die, they explode with such immense force that material is pushed inwards as well as exploding outwards. The inward-rushing mass becomes so dense that it collapses under its own gravity until it suddenly blinks out of the realm of normal time and space, becoming a dimensionless point of infinitely condensed matter with a truly monstrous gravitational field. Nothing can escape this field. Not even light. Hence the term 'black hole'.

Black holes are therefore the ultimate challenge for science, because they are objects which, by their very nature, can never be directly seen through our telescopes or measured with instruments. (In the British TV comedy series *Red Dwarf*, a talking computer called Holly gives a grudging apology for nearly crashing her crew into one of these monsters: 'The background colour of space is basically black. And black holes – well, they're black. So they're a bit hard to spot. Sorry.') We can only infer their presence from circumstantial evidence: stars that appear to be under vicious attack from some nearby invisible entity, or the unusually powerful burst of x-rays given off as the very atoms of gas and dust material are torn apart in the final nanoseconds before disappearing into a black hole.

The centres of galaxies are more densely packed with stars than the perimeters, and new black holes are more likely to form there because of the relative abundance of material to feed upon. However, some theories suggest that black holes also have a useful aspect: they are a centre of mass around which healthy young galaxies can coalesce.

Many cosmologists suggest that when a black hole reaches its point of infinite compression, it creates an equivalent 'white hole' in another universe, spewing out matter in a Big Bang. Our universe emerged suddenly from a dimensionless point of infinitely compressed matter. A black hole is ... a dimensionless point of infinitely compressed matter. To hijack a quote from Bart Simpson when confronted by a malfunctioning vacuum cleaner: 'This does something I once thought was impossible. It both sucks and blows.' Perhaps a black hole in an earlier universe gave birth to ours, while black holes in our universe are spawning yet more.

But surely a black hole would have to swallow an entire universe in one set of time-space dimensions before it could spit it out into another? Not necessarily. A black hole can swallow a few stars, and perhaps just the remnants of the one star it first came from, into a pinprick of infinitely condensed matter. Yet it takes only one such pinprick to make an entire universe.

If black holes are capable of creating new universes, this might explain how our particular universe could have arisen from a process quite similar to Darwinian evolution. In November 1996, two biologists, John Maynard Smith of the University of Sussex, and Eörs Szathmary of the Collegium Budapest in Hungary, wrote an informal essay for the science journal *Nature*, entitled 'On the Likelihood of Habitable Worlds'. They adapted a number of perfectly respectable scientific concepts in support of an amazing idea of their own. This is how it works:

Our universe must be well-tuned for life because we are here to discuss it. Our universe also makes plenty of black holes, as a result of the same laws of physics (the four fundamental forces) that make life possible. Those black holes, in turn, are spewing out plenty of other new universes. So, a universe that's good for black holes, and also suitable for life, is more likely to create new universes.

On the other hand, a universe that's *not* suitable for life is also not so likely to create black holes. So it won't generate other universes. The more new universes that are created, the higher the chances that some of them will create their own black holes. And so on. The process is exponential, and life-suited universes will tend to predominate. And so, the 'hand-set values' dilemma in the strong anthropic principle is reversed. Instead of being a one-off fluke, it is almost inevitable that we should find ourselves in a universe with a life-friendly set of physical laws and a stable balance between the four fundamental forces.

To recap: the anthropic principle invites us to hold two opposed but equally mind-boggling possibilities in our heads at the same time.

1) There is only one universe, and against colossal odds it is exceptionally well-adjusted for complex things like stars, planets and living creatures to emerge. It looks almost as if it has been designed that way.

2) There are billions upon billions of universes out there in a great 'multiverse' spawned by black hole/white hole singularities. These universes can be long-lasting, short-lived, stable, unstable, big and small, hot or cold, poisonous

or paradisiacal, but universes like ours will tend to predominate because they are more likely to create black holes, and therefore to seed more universes. So it's no great surprise to find ourselves here. The multiverse has already thrown the Big Bang dice billions of times, and there was no need for any Designer to stack the odds in our favour.

Cosmologist Lee Smolin has shown that it might be possible, in the coming decade or two, to make specific astronomical observations in support of the black hole/white hole theory. We might soon be able to prove that multiple universes actually exist. We'll need a new generation of powerful telescopes. They are on the space planners' digital drawing boards even now. In the meantime, the debate about multiple universes continues. This is no science fiction backwater. It is a central question about the nature of reality that must be resolved. In April 2003, the *New York Times* ran an article by the British physicist Paul Davies, warning readers not to take multiple universes too seriously. He argued that it's not good science to speculate about things that are intrinsically unobservable, because there's no way to prove them. The next month, *Scientific American* published an article by physicist Max Tegmark asserting that parallel universes almost certainly *must* exist. On the other hand, the late Martin Gardner, renowned for his mathematical inventiveness, said: 'Surely the conjecture that there is just one universe and its Creator is infinitely simpler and easier to believe than that there are countless billions upon billions of worlds?'

Are we faced, in the meantime, with a straight choice between one designer-built universe, or billions of them thrown out by pure chance? Not if we think beyond the surface dazzle of stars and galaxies and look deeper at the possible underlying nature of the multiverse.

The matrix

Physics seems to be showing us that everything we know of, from cats and dogs and trees and people, to stars, planets and galaxies, is all made from atoms and subatomic particles: in other words, from a relatively small kit of fundamental entities out of which the sum total of existence is made. We may be closing in on a set of equations that can account for everything using a few simple rules. Scientists argue vigorously about the details, but there is a marked tendency for most of them to believe that some truly simple building block, even simpler than the subatomic components we've so far explored, accounts for all the universe's matter and energy. Those strings, maybe. We might then be left with a more subtle question. How do we account for the *differences* we observe between things, when they are all made of the same kind of stuff? One possible answer is that the cosmos behaves like a computer, 'outputting' the version of reality that we observe, rather than actually *being* that reality. It uses a code of subatomic components and forces, and combines them according to the rules of its software – what we think of as 'the laws of physics' – to deliver many different outputs, just as an electronic computer manipulates simple on-off pulses, noughts and ones, to create apparently complex three-dimensional images on a virtual reality screen. Scientists are now debating the potential importance of *information* as the true mechanism at the heart of existence. Any hydrogen atom or oxygen atom or electron or proton or whatever is exactly the same (apart from isotopic variations) as any of its cousins, no matter that there are countless trillions of them in the universe as a whole. They're just pieces of information, bits of code.

Pull back to the macroscopic scale, though, and distinct chemical structures begin to emerge. Pull back further, and the cosmos seems suddenly rich in its variety of 'things' like galaxies, stars, planets, cats, dogs and people (and not forgetting, as the *Hitchhiker's Guide to the Galaxy* cruelly reminds us, 'cheap digital watches'). Yet these apparently solid constructions may all be nothing more than shimmering assemblages of subatomic information, as malleable and essentially impermanent as any computer's temporarily stored patterns of digital data, brought briefly to apparent life on a plasma screen. Imagine that screen not as a flat surface on your laptop, but as something multi-dimensional, and you have some flavour of how the universe as we *think* we see and touch and feel it might be a kind of virtual construct.

One key point about a computer is that the information inside it has to be manipulated by some kind of surrounding physical architecture: a modern computer chip, for instance, with its millions of transistor switches and memory modules for storing the results of its calculations. There is some confusion among information theorists about whether or not information has any real meaning in the absence of such a system to process it, and just as importantly, something – or someone – to *read* it. If the cosmos is similarly information-based, what is the hidden architecture that stores all the information, manipulates it and outputs the results? And who's watching the output, interpreting it and judging it to have some kind of meaning?

From our point of view, observing atomic and subatomic particles of matter and energy with our scientific instruments is roughly the equivalent of gazing through a powerful magnifying glass at the tiny pixels on the screen of a vast computer. We can learn a lot about the wonderful illusions that appear on the screen when all the pixels combine to make real-seeming pictures. But we have to understand the processing architecture *behind* the screen if we are really to find how the cosmos creates those illusions.

This idea might seem far-fetched, like something out of the movie *The Matrix*, until we remember that we have already learned to simulate most kinds of familiar visual and sound experiences using nothing more than simple binary code computers. In all likelihood we are less than a couple of decades away from creating illusions indistinguishable from our current notions of reality. We might then be able to step outside our arcade simulators at the end of a day's play and re-enter what we think of as the 'real' world, but there may be no such obvious exit from the greater cosmic illusion all around us.

As we've seen, the first generation of stars was made entirely of helium and hydrogen, and nothing else. Nuclear fission inside some of those stars manufactured all the more complicated chemical elements (atoms) that we are familiar with: oxygen, carbon, nitrogen, iron, magnesium and so on. We are all star stuff. We have borrowed them, and they are already billions of years old. After our bodies have finished with them, those atoms will survive for many more billions of years. Living creatures such as ourselves could be thought of as fleeting patterns made from these long-lived bits of information. What, then, makes us think we are 'real'?

It's worth reminding ourselves that computers use just two kinds of information to create complex patterns and illusions. Our DNA uses four kinds of coding, called base pairs, and the entire universe (including DNA and computers) is built from atoms, which as we've seen are extremely standardised, and which (as Rutherford proved) consist almost entirely of the empty space between the electrons and the nucleus. There is a good argument to suggest that information is more important than solid 'stuff' in shaping reality.

Here's a puzzle. Lego bricks are famous for the way they can be stacked together into a limitless variety of shapes. You can build an aeroplane, a spaceship or a house, yet the bricks are rather like atoms and subatomic particles. They are indistinguishable from each other, and they come in a limited 'family' of types. Let's imagine a Lego kit with just three kinds of brick, representing the three basic components of all atoms: the 'up' and 'down' quarks that make up atomic nuclei and the electrons that go round the outside. Now, suppose you build a beautiful life-size cathedral, using tens of thousands of very small bricks, in just these three varieties, to build up the finest details. And you live a long life, then die, and the model passes to your children. Then it languishes in a museum for hundreds of years. Finally, somehow, it survives long after all human culture has vanished in a terrible calamity, and there's no one left to admire it, or to understand its construction or what it originally meant when you first built it all those aeons ago. Is it still a beautiful model of a cathedral or just a pile of Lego bricks?

Does the universe 'exist'? Or do conscious creatures such as ourselves need to be aware of it to give it meaning above and beyond just its atomic building bricks? Again, if this sounds like an outlandishly weird question, it's just a simple description of a problem that has vexed the quantum science community ever since the 1927 Solvay Conference. Do we, as conscious observers, actually *make* the universe real by interpreting the information it's made of and giving it meaning? According to physicist Paul Davies, 'through conscious beings, the universe has generated self-awareness. This can be no trivial detail, no minor by-product of mindless, purposeless forces. We are truly meant to be here.'

The scientific viewpoint, of necessity, is an invention of our animal minds. It may be that an animal mind is not capable of understanding anything beyond the animal circumstance. We human animals try to discover the mysteries of the universe with minds adapted for survival on earth, and therefore ill-adapted for speculations about time, space, matter and energy. The fact that we've been privileged to gain *any* insights into the deeper meanings of it all is remarkable.

In 1984, Peter Medawar, a respected biologist, wrote an essay entitled 'The Limits of Science'. He warned that scientists were not entitled to say that they could explain everything in terms of scientific constructions, atoms, electrons, gravity fields and so forth, which could not themselves be explained. With crisp logic, he argued that science can answer only scientific questions, and many questions are just *not* scientific. 'There are questions children ask. How did everything begin? What are we all here for? What is the point of life? It would be universally agreed that it is not to science that we should look for answers to those questions. There is, then, a limit to scientific understanding.' One thing's for sure. That limit has not yet been reached. Who knows what new and extraordinary things we will discover in the coming generation? The atom has sent us on a journey that is far from over.

First and last things

By weight, our bodies are 10 per cent hydrogen, 60 per cent oxygen and 20 per cent carbon. The last 10 per cent is taken up principally by nitrogen, calcium, phosphorous, sulphur, sodium and magnesium, iron and copper. The atoms that constitute these substances were created not in our sun but by other stars long since dead, as we have seen. They are not our atoms. They do not belong to us, nor to our bodies. They are just passing through. Every moment of our lives, we exhale carbon dioxide. The atoms in that gas have existed for billions of years already, and they will continue to exist for billions of years more after we are gone; in soil, in leaves, in dogs, in cats, in flowers, and in the air breathed by our descendants.

Our sun has burned for five billion years, and will burn for five billion more; after which, it will have used up a crucial proportion of its available hydrogen. The sun that has nurtured us for so long will turn into a violent and unpredictable monster prone to sudden nuclear shutdowns and re-ignitions, expansions and contractions, alternating over several millennia. It will end its days as a red giant so gigantic that its wayward mantle of gases will swallow up half the planets of the solar system, including the earth. Everything on our planet that lives and breathes will be utterly vaporised. The grass, the trees, even the toughened lichen that clings to granite will be blasted into atoms. The oceans will boil away to the last drop. The ice caps will vaporise. The sands on every beach in the world will fuse into glass. Mercury and Venus will vanish. If earth survives the storm, it will remain only as a scorched lump of rock. The sun's final life-giving service will have been to vaporise the thin sliver of biosphere on the earth's surface, blasting it into space once more so that the atoms might one day contribute towards new worlds and new forms of life. Perhaps, some day in the far future, some intelligent and curious creature containing a few of our second-hand atoms will ask where they came from.

Further Reading

David Bodanis, *E = mc²: A Biography of the World's Most Famous Equation*, Macmillan, London, 2000

Edmund Blair Bolles, *Einstein Defiant: Genius versus Genius in the Quantum Revolution*, Joseph Henry Press, Washington DC, 2004

Melvyn Bragg, *On Giants' Shoulders: Great Scientists and Their Discoveries from Archimedes to DNA*, Hodder and Stoughton, London, 1998

David C. Cassidy, *Uncertainty: The Life and Science of Werner Heisenberg*, W.H. Freeman, New York, 1991

Brian Cathcart, *The Fly in the Cathedral: How a Small Group of Cambridge Scientists Won the Race to Split the Atom*, Penguin Books, London, 2005

Marcus Chown, *Afterglow of Creation: From the Fireball to the Discovery of Cosmic Ripples*, Arrow Books, London, 1993

Barbara Lovett Cline, *Men Who Made a New Physics: Physicists and the Quantum Theory*, University of Chicago Press, 1965

William H. Cropper, *Great Physicists: The Life and Times of Leading Physicists from Galileo to Hawking*, Oxford University Press, 2001

Nuel Pharr Davis, *Lawrence and Oppenheimer*, Simon and Schuster, New York, 1968

Richard P. Feynman, *'What Do You Care What Other People Think?' Further Adventures of a Curious Character*, Unwin Hyman Limited, London, 1989

James Gleick, *Genius: Richard Feynman and Modern Physics*, Abacus, London, 1994

Bruce Gregory, *Inventing Reality: Physics as Language*, Wiley Science Editions, New York, 1988

Tony Hey and Patrick Walters, *The New Quantum Universe*, Cambridge University Press, 2003

J.P. McEvoy and Oscar Zarate, *Introducing Quantum Theory*, Icon Books, Cambridge, 2004

Jean Medawar and David Pyke, *Hitler's Gift: Scientists Who Fled Nazi Germany*, Richard Cohen Books, London, 2000

Ruth Moore, *Niels Bohr, the Man and the Scientist*, Hodder and Stoughton, London, 1967

Walter Moore, *A Life of Erwin Schrödinger*, Cambridge University Press, 1994

Ed Regis, *Who Got Einstein's Office? Eccentricity and Genius at the Princeton Institute for Advanced Study*, Penguin Books, London, 1989

Index